Low Cost Waste water Treatment Technologies

Low Cost Waste water Treatment Technologies

Mariyam Noor

CYBER TECH PUBLICATIONS
4264/3, Ansari Road, Daryaganj, New Delhi-110002 (India)
Ph.: 011-23244078, 011-43559448 Fax: 011-23280028
E-Mail: cyberpublicationsdelhi@yahoo.com
Website: www.cybertechpublications.com

LOW COST WASTE WATER TREATMENT TECHNOLOGIES

Mariyam Noor

First Edition 2011

Published by :

G.S. Rawat for **Cyber Tech Publications**
4264/3, Ansari Road, Daryaganj, New Delhi-110002 (India)
Ph.: 011-23244078, 011-43559448 Fax: 011-23280028
E-Mail: cyberpublicationsdelhi@yahoo.com
Website: www.cybertechpublications.com

Printed at: Asian Offset Printers, Delhi-53

CONTENTS

Preface

The sustainable use of coastal and ocean resources is linked to public health, food security, and economic and social benefits, including cultural values and traditional livelihoods. More broadly, these issues are understood as decisive elements in the alleviation of poverty. Ironically, the major threat to the health, productivity and biodiversity of the world's oceans is human activity on land - both in coastal areas and much further inland away from the sea.

If adequate measures are not taken, rapid urbanisation and industrialisation, especially in coastal areas, can adversely impact the fragile balance of estuarine, near-shore and marine ecosystems. In particular, uncontrolled discharge of municipal sewage is one of the most serious forms of ocean pollution, affecting every region of the world. A majority of domestic sewage in developing countries is discharged directly into local waterways, international rivers or the sea without treatment. This is still the case even in some developed countries.

In response to the daunting challenge faced by many governments in addressing the problems outlined above, the subject document was prepared as a collaborative venture of the United Nations Environment Programme (UNEP), the World Health Organisation (WHO), the United Nations Human Settlements Programme (UN-HABITAT) and the Water Supply and Sanitation Collaborative Council (WSSCC). It provides guidance for local and national policy and decision-makers, as well as professionals, on the identification, selection and application of appropriate, environmentally sound and low cost systems for municipal wastewater management.

—*Author*

1

Introduction and Overview

An interesting observation arising from the preparation of this year's special chapter on water and agriculture is how difficult it is to generalize about water. Almost any statement requires qualification. For example, while we can say that water is one of the most abundant resources on earth, we know that less than 1 percent of the total supply is reliably available for human consumption. Water is a liquid for the most part, but it can also be a solid and a vapour. Drinking-water is certainly essential for human survival but water-related illnesses are the most common health threat in the developing world. An estimated 25 000 people die every day as a result of water-related sicknesses. One statement, however, needs no qualification: human existence depends on water. The geosphere, the atmosphere and the biosphere are all linked to water. Water interacts with solar energy to determine climate and it transforms and transports the physical and chemical substances necessary for all life on earth.

In recent years, water issues have been the focus of increasing international concern and debate. From 26 to 31 January 1992, the UN system sponsored the International Conference on Water and the Environment (ICWE) in Dublin, Ireland. The ICWE called for innovative approaches to the assessment, development and management of freshwater resources. In addition, the ICWE provided policy guidance for the United Nations Conference on Environment and Development (UNCED) in Rio de Janeiro in June 1992. UNCED highlighted the need for water sector reforms throughout the world.

In 1993, the World Bank issued a comprehensive policy paper defining its new objectives for the water sector. FAO recently established an International Action Programme on Water and Sustainable Agricultural Development (IAP-WASAD). Likewise, the UNDP, WHO, UNICEF, WMO, Unesco and UNEP are all coordinating or participating in special programmes related to water resources.

Other international, national and local organizations are becoming more active in water issues. The 1990 Montreal meeting, "NGOs Working Together", focused attention on drinking-water supply and sanitation. The Canadian International Development Agency, the French Ministry of Cooperation and Development, the German Agency for Technical Cooperation (GTZ), the United Kingdom's Overseas Development Administration and the United States Agency for International Development (USAID) have recently developed water resource strategies for foreign assistance.

The message highlighted by all these efforts is that water is an increasingly scarce and valuable resource. Of principal concern is our failure to recognize and accept that there is a finite supply of water. The consensus is that the growing water scarcity and misuse of freshwater pose serious threats to sustainable development.

Competition among agriculture, industry and cities for limited water supplies is already constraining development efforts in many countries. As populations expand and economies grow, the competition for limited supplies will intensify and so will conflicts among water users.

Despite water shortages, misuse of water is widespread. Small communities and large cities, farmers and industries, developing countries and industrialized economies are all mismanaging water resources. Surface water quality is deteriorating in key basins from urban and industrial wastes.

Groundwater is polluted from surface sources and irreversibly damaged by the intrusion of salt water. Overexploited aquifers are losing their capacity to hold water and lands are subsiding. Cities are unable to provide adequate drinking-water and sanitation facilities. Waterlogging and salinization are diminishing the productivity of irrigated lands. Decreasing water flows are reducing hydroelectric power generation, pollution assimilation and fish and wildlife habitats.

At first glance, most of these water problems do not appear to be directly related to the agricultural sector. Yet, by far the largest demand for the world's water comes from agriculture. More than two-thirds of the water withdrawn from the earth's rivers, lakes and aquifers is used for irrigation. As competition, conflicts, shortages, waste, overuse and degradation of water resources grow, policy-makers look increasingly to agriculture as the system's safety valve.

Agriculture is not only the world's largest water user in terms of volume, it is also a relatively low-value, low-efficiency and highly subsidized water user. These facts are forcing governments and donors to rethink the economic, social and environmental implications of large publicly funded and operated irrigation projects. In the past, domestic spending for irrigation dominated agricultural budgets in countries throughout the world. For instance, since 1940, 80 percent of Mexico's public expenditures in agriculture have been for irrigation projects. In China, Indonesia and Pakistan, irrigation has absorbed more than half of agricultural investment. In India, about 30 percent of all public investment has gone into irrigation.

A significant portion of international development assistance has also been used to establish irrigation systems. Irrigation received nearly 30 percent of World Bank agricultural lending during the 1980s. Spending commitments for irrigation by all aid agencies exceeded $2 billion per year in the past decade.

Once established, irrigation projects become some of the most heavily subsidized economic activities in the world. In the mid-1980s, Repetto estimated that average subsidies to irrigation in six Asian countries covered 90 percent of the total operating and maintenance costs. Case-studies indicate that irrigation fees are, on average, less than 8 percent of the value of benefits derived from irrigation.

Despite these huge investments and subsidies, irrigation performance indicators are falling short of expectations for yield increases, area irrigated and technical efficiency in water use. As much as 60 percent of the water diverted or pumped for irrigation is wasted. Although some losses are inevitable, in too many cases this excess water seeps back into the ground, causing waterlogging and salinity. As much as one-quarter of all irrigated land in developing countries suffers from varying degrees of salinization. Moreover, stagnant water and poor irrigation drainage escalate the incidence of water-related diseases, resulting in human suffering and increased health costs.

Today, agriculture is often unable to compete economically for scarce water. Cities and industries can afford to pay more for water and earn a higher economic rate of return from a unit of water than does agriculture. (For economists, water flows uphill to money.) For the first time in many countries, agriculture is being obliged to give up water for higher-value uses in cities and industries. Irrigators in some areas are now asked to pay for the water they receive, including the full cost of water delivery. In other areas, new regulations require farmers to pay for polluting streams, lakes and aquifers.

The irony is that irrigated agriculture is expected to produce much more in the future while using less water than it uses today. At present, 2.4 billion people depend on irrigated agriculture for jobs, food and income (some 55 percent of all wheat and rice output is irrigated). Over the next 30 years, an estimated 80 percent of the additional food supplies required to feed the world will depend on irrigation.

These developments are placing enormous pressure on agricultural policy-makers and farmers. Throughout the world, governments assume the prime responsibility for ensuring food security and, because food depends increasingly on irrigation, food security is closely linked with water security. Between 30 and 40 percent of the world's food comes from the irrigated 16 percent of the total cultivated land; around one-fifth of the total value of fish production comes from freshwater aquaculture; and current global livestock drinking-water requirements are 60 billion litres per day (forecasts estimate an increase of 0.4 billion litres per year). Food security in the next century will be closely allied to success in irrigation. Irrigation can help make yield-increasing innovations a more attractive investment proposition but it does not guarantee crop yield increases. The overall performance of many irrigation projects has been disappointing because of poor scheme conception, inadequate construction and implementation or ineffective management. The mediocre performance of the irrigation sector is also contributing to many socio-economic and environmental problems, but these problems are neither inherent in the technology nor inevitable, as is sometimes argued.

Irrigation projects can contribute greatly to increased incomes and agricultural production compared with rain-fed agriculture. In addition, irrigation is more reliable and allows for a wider and more diversified choice of cropping patterns as well as the production of higher-value crops. Irrigation's contribution to food security in China, Egypt, India, Morocco and Pakistan is widely recognized. For example, in India, 55 percent of agricultural output is from irrigated land. Moreover, average farm incomes have increased from 80 to 100 percent as a result of irrigation, while yields have doubled compared with those achieved under the former rain-fed conditions; incremental labour days used per hectare have increased by 50 to 100 percent. In Mexico, half the value of agriculture production and two-thirds of the value of agricultural exports is from the one-third of arable land that is irrigated.

Irrigation is a key component of the technical package needed to achieve productivity gains. In the future, as high levels of costly inputs are added to cropland to sustain yield increases, the security and efficiency of irrigated production will become even more important to world farming. Water will no longer be plentiful and cheap. It will be scarce, expensive to develop and maintain and valuable in use. The prospect of high-cost water may at first seem to be another problem looming for low-income economies. However, the high cost will be an incentive to use water more efficiently. The single most important factor limiting the adoption of proven irrigation and drainage technology is the low cost of water. Moreover, if farmers have opportunities for higher-value uses and can make profits, both governments and farmers will invest in irrigation. This water dilemma-to produce more in a sustainable way with less water-points to the need for demand management mechanisms to reallocate existing supplies, encourage more efficient use and promote more equitable access. Policy-makers need to establish a structure of incentives, regulations, permits, restrictions and penalties that will help guide, influence and coordinate how people use water while encouraging innovations in water-saving technologies.

In the past, supply-side approaches dominated water resource management practices. Water itself was physically managed through technical and engineering means that captured, stored, delivered and treated water. However, the era of meeting growing demand by developing new supplies is ending. In our present-day water economy, resource management is shifting away from the goal of capturing more water towards that of designing demand-and user-focused approaches that influence behaviour.

PURPOSE AND SCOPE

This special chapter is primarily intended for agricultural policy-makers, water managers, researchers, students, development planners and agricultural project donors. It is meant to help us reflect on the way water resources are managed at present; to contribute to the discussion on sustainable water use; and to stimulate thinking, research and change. Decisions made in this decade regarding how water is used will have a profound effect on our future supplies.

This first section gives an overview of world water resources and briefly discusses the key issues: scarcity, quality and health.

The second section stresses the need to integrate the water sector with the national economy and analyses the physical, economic and social aspects of water. It then provides a conceptual foundation for understanding the circumstances under which water policies either work or fail. Section II also assesses the advantages and disadvantages of broad alternative approaches to public water policy. Section III examines how policy analysis is applied to water resource planning, including both supply-side (physical and hydrological) and demand-side considerations. It discusses the advantages and disadvantages of various policy options for urgent water policy issues related to surface water and groundwater.

The fourth and final section reviews three specific policy issues in irrigated agriculture: declining growth and investment trends; the difficulties imposed by irrigation-induced environmental degradation; and efforts to reform managerial and administrative systems.

WORLD WATER RESOURCES

Every day the hydrological cycle renews the world's freshwater resources through evaporation and precipitation. The average annual rainfall over land is 110 000 km^3, but some 70 000 km3 evaporate before reaching the sea. The remaining 40 000 km3 are potentially available for human use. Global freshwater consumption is currently around 4 000 km3, only 10 percent of the annual renewable supply.

These numbers suggest that plenty of water is available for human use but a closer look reveals a more complicated situation. The 40 000 km^3 of available water are distributed very unevenly and two-thirds of it runs off in floods. That leaves around 14 000 km^3 as a relatively stable supply. A substantial share of this supply should be left to follow its natural course in order to safeguard wetlands, deltas, lakes and rivers.

For example, 6 000 km^3 of water is needed to dilute and transport the estimated 450 km^3 of waste water now entering the world's rivers each year. Without substantial investment in waste water treatment and more effective regulation, even more water will have to be diverted to dilute and transport wastes.

Precipitation, withdrawals and availability of water vary widely around the world. Per caput availability is highest in Latin America and lowest in North Africa and the Near East, while withdrawals are highest in North America and lowest in Africa. Per caput water availability in Europe and North America is not expected to change greatly by 2000, while Asians, Africans and Latin Americans will face less per caput water availability as their populations continue to grow.

At present, Asia accounts for over one-half of the world's water withdrawals. Forecasts to the year 2000 suggest that Asia will consume 60 percent of the world's water, followed by 15 percent in North America, 13 percent in Europe and less than 7 percent in Africa. Latin America's share of world water consumption is forecast to be less than 5 percent in 2000, although the region's consumption has nearly quadrupled since 1950.

WATER SCARCITY

Human actions bring about water scarcity in three ways: through population growth, misuse and inequitable access. Population growth contributes to scarcity simply because the available water supply must be divided among more and more people. Every country has a more or less fixed amount of internal water resources, defined as the average annual flow of rivers and aquifers generated from precipitation. Overtime, this internal renewable supply must be divided among more and more people, eventually resulting in water scarcity.

TABLE : Per caput water availability by region, 1950-2000

Region	1950	1960	1970	1980	2000
	(........................		'000 m3		........................)
Africa	20.6	16.5	12.7	9.4	5.1
Asia	9.6	7.9	6.1	5.1	3.3
Latin America	105.0	80.2	61.7	48.8	28.3
Europe	5.9	5.4	4.9	4.4	4.1
North America	37.2	30.2	25.2	21.3	17.5

Source: N.B. Ayibotele. 1992. The world's water: assessing the resource. Keynote paper at the ICWE, Dublin, Ireland.

When annual internal renewable water resources are less than 1 000 m^3 per caput, water availability is considered a severe constraint on socio-economic development and environmental protection. Table 7 lists the countries where per caput internal renewable water availability will fall below 1 000 m^3 by the end of this decade. Most countries facing chronic water scarcity problems are in North Africa, the Near East and sub-Saharan Africa. Countries with less than 2 000 m^3 per caput face a serious marginal water scarcity situation, with major problems occurring in drought years. By the end of the 1990s, water availability is expected to fall below 2 000m^3 per caput in more than 40 countries.

In many countries, while scarcity is less of a problem at a national level, serious water shortages are causing difficulties in specific regions and watersheds. Notable examples include northern China, western and southern India and parts of Mexico.

TABLE : Countries predicted to have scarce water resources in 2000

Country[1]	Population in 2000	Water availability	
		Internal renewable water resources	Water resources including river flows from other countries
	(millions)	*(....m^3 per caput....)*	
Egypt	62.4	29	934
Saudi Arabia	21.3	103	103
Libyan Arab Jamahiriya	6.5	108	108
United Arab Emirates	2.0	152	152
Jordan	4.6	153	240
Mauritania	2.6	154	2843
Yemen	16.2	155	155
Israel	6.4	260	335
Tunisia	9.8	384	445
Syrian Arab Republic	17.7	430	2008
Kenya	34.0	436	436
Burundi	7.4	487	487
Algeria	33.1	570	576
Hungary	10.1	591	11 326
Rwanda	10.4	604	604
Botswana	1.6	622	11 187
Malawi	11.8	760	760
Oman	2.3	880	880
Sudan	33.1	905	3923
Morocco	31.8	943	943
Somalia	10.6	1 086	1 086

People also bring about water scarcity by polluting and overusing existing supplies. This type of scarcity can be regarded as the consumption of the resource's "capital". For instance, an aquifer represents resource capital, providing what is generally a renewable source of water "income" that can be tapped for human consumption. Sustainable use of the aquifer leaves the capital intact so that future generations can continuously use the renewable portion or income. If pumping is greater than recharge, the aquifer is depleted and the capital is consumed.

Table : Ratio of prices charged by vendors to prices charged by public utilities in selected cities

Country	City	Ratio
Bangladesh	Dacca	12-25
Colombia	Cali	10
Côte d'Ivoire	Abidjan	5
Ecuador	Guayaquil	20
Haiti	Port-au-Prince	17-100
Honduras	Tegucigalpa	16-34
Indonesia	Jakarta	4-60
	Surabaja	20-60
Kenya	Nairobi	7-11
Mauritania	Nouakchott	100
Nigeria	Lagos	4-10
	Onitsha	6-38
Pakistan	Karachi	28-83
Peru	Lima	17
Togo	Lomé	7-10
Turkey	Istanbul	10
Uganda	Kampala	4-9

Source: R. Bhatia and M. Falkenmark. 1992. Water resource policies and the urban poor: innovative approaches and policy imperatives. Background paper for the ICWE, Dublin, Ireland.

Overuse of groundwater has become a major problem in China, India, Indonesia, Mexico, the Near East, North Africa, Thailand, the western United States and many island countries where seawater intrusion results. The overpumping of aquifers not only results in a water source that is too depleted to serve as a supply, it may also cause the land above the aquifer to settle or subside, resulting in widespread structural damage in extreme cases. Bangkok and Mexico City are well-known examples.

Finally, a shift in access or distribution patterns may concentrate water resources among one group and subject others to extreme scarcity. In many cities of the developing world, large numbers of people depend on water vendors and may pay 100 times as much as the rate of public utilities. Numerous recent studies document that large numbers of urban poor pay much higher prices and a much larger share of their income for water than families with access to a city water system. Poor families in some large cities spend up to 20 percent of their income on water. When the cost is so high, they use little water for washing and bathing, which results in serious health problems.

A number of other countries with smaller populations, e.g. Barbados, Cape Verde, Djibouti, Malta, Qatar and Singapore, are also included in the water-scarce category. Source: FAO calculations based on World Bank/WRI data.

World water use

The early civilizations of Asia, Africa and Latin America organized cooperative efforts to develop river valleys for irrigated agriculture. Through irrigation technology, societies controlled and manipulated natural water supplies to improve crop production. The result was often reliable and ample food supplies which led to the creation of stable agricultural villages, the division of labour and economic surpluses.

Many scholars still argue over whether irrigation technology facilitated political control and development of the state or whether political developments led to advancement of the technology. No matter the direction of cause and effect, no one disputes the association of development with control over water use.

In today's world, agriculture still accounts for the majority of human water use. Globally, around 70 percent of water withdrawals are for agriculture. Domestic and industrial uses consume the remaining 30 percent. Domestic uses include drinking-water supplies, private homes, commercial establishments, public services and municipal supplies.

Water uses differ greatly depending on access, quantity, quality and socio-economic conditions. For example, agricultural water use is higher as a proportion of total water use in the low-income countries (91 percent) than in the high-income group (39 percent).

TABLE : Sectoral water withdrawals, by income group

Country income group	Annual withdrawals per caput (...m³...)	Withdrawals by sector Agric.	Ind.	Dom.
		(...............%..........)		
Low-income	386	91	5	4
Middle-income	453	69	18	13
High-income	1 167	39	47	14

Source: World Bank. 1992. World Development Report 1992, based on WRI data.

Nevertheless, on a per caput basis, the high-income countries use more water for agricultural purposes than the low-income countries. The trends in world water use during this century are presented in Figure 12. Overall, global water consumption has increased almost tenfold. Agriculture's share, which was 90 percent in 1900, will have dropped to an estimated 62 percent by 2000.

During this same period, industrial consumption will have grown from 6 percent to 25 percent, while consumption by cities will have increased from 2 percent to nearly 9 percent. By 2000, around 35 percent of available water supplies will be in use, compared with less than 5 percent at the beginning of the century.

Water quantity and quality requirements also differ widely depending on the type of use. Net agricultural requirements are especially large in relation to other uses. For instance, around 15 000 m3 of water are normally sufficient to irrigate 1 ha of rice. This same amount of water can supply: 100 nomads and 450 head of stock for three years; or 100 rural families through house connections for four years; or 100 urban families for two years; or 100 luxury hotel guests for 55 days.

WATER AND POLLUTION

The quality of water from different sources varies widely. Precipitation absorbs gases from the atmosphere and removes particles from the air. When the precipitation strikes the ground it becomes surface water runoff or enters the ground. The surface water flows into larger and larger channels, ponds, lakes and rivers until some of it reaches the sea. Along its course, surface water picks up both organic and mineral particles, bacteria and other organisms as well as salts and other soluble substances. The water in lakes and swamps sometimes acquires odours, tastes and colours from algae and other organisms and from decaying vegetation.

Since ancient times, heavy metals from mining and pathogens from cities have caused serious, although localized, contamination. Since the industrial revolution, water pollution problems have become first regional, then continental and now global in nature. Much water is polluted when it is used in industry and agriculture or for domestic purposes. Mining is the major cause of metal contamination, whereas other industries contribute to acidification. The intensification of agricultural activities has led to the contamination of groundwater by fertilizers and other chemicals. Moreover, irrigation projects often cause a rapid rise in the level of groundwater, which leads to waterlogging and soil salinity.

Since 1977, the UNEP/ WHO Global Environment Monitoring System (GEMS) has been working with Unesco and WMO to develop a global water quality monitoring network. More than 50 water variables are monitored to provide information on the suitability of water for human consumption and for agricultural, commercial and industrial use. Recent assessments have found that the main water pollutants are: sewage, nutrients, toxic metals and industrial as well as agricultural chemicals.

Conclusions drawn from the GEMS assessment include: the nature and level of freshwater pollution strongly depends on socio-economic development; the most common water pollutant is organic material from domestic sewage, municipal waste and agro-industrial effluent; and the high water nitrate levels found in Western Europe and the United States are a result of the nitrogen fertilizers and manure used for intensive agriculture. The GEMS assessment also noted a dramatic increase in the use of fertilizers in developing countries, particularly where intensive irrigation allows for double or triple cropping.

Other conditions highlighted in the GEMS report include deforestation, eutrophication, suspended particulate matter (SM) and salinity.

Deforestation, i.e. the clearing of land for agriculture and urban development, often leads to water contamination. When the soil is stripped of its protective vegetative covering, it becomes prone to erosion. This in turn leads to higher water turbidity, because of the increased amounts of suspended matter, to nutrient leaching and to a decreased water-retention capacity of the soil. There is also concern about the destruction of wetlands, which destroys the habitat of many species and removes natural filter mechanisms, permitting many common pollutants to reach water supplies.

Eutrophication is the enrichment of waters with nutrients, especially phosphorus and nitrogen. It can lead to enhanced plant growth and depleted oxygen levels as this plant material decays. It is not always a human-induced problem, but is often linked to organic waste and agricultural runoff. Today 30 to 40 percent of the world's lakes and reservoirs are eutrophic. Not all interventions have proved successful, but eutrophication can be reversible if mid-and long-term strategies are enacted. Laws and measures introduced to reduce tripolyphosphates (used mostly in detergents) and to remove phosphorus from waste water have had positive effects.

SM consists of materials that float in suspension in water. There are three main sources of SM: natural soil erosion, matter formed organically within a water body and material produced as a by-product of human activity. SM settles on the sediment bed and forms deposits in rivers, lakes, deltas and estuaries.

Evidence of human-induced SM from Roman and Mayan times has been discovered in lake beds, implying that this was one of the first types of water pollution. River damming affects the amount of SM flowing from rivers to the oceans because reservoirs act as effective sinks for SM. An estimated 10 percent of the global SM discharge to the sea is trapped in reservoirs. Approximately 25 percent of the water currently flowing to the oceans has been previously stored in a reservoir. Damming can also greatly modify water quality; waters flowing out of reservoirs not only have reduced SM quantities, they are also depleted of nutrients and are often more saline, which consequently has detrimental effects on downstream agriculture and fisheries.

Salinity is a significant and widespread form of freshwater pollution, particularly in arid, semi-arid and some coastal regions. The primary cause of salinization is a combination of poor drainage and high evaporation rates which concentrate salts on irrigated land. Salinity can adversely affect the productivity of irrigated crops and is also detrimental to industrial and household water users. It is not a new phenomenon; salinization of soil and water in the flood plain of the Tigris and Euphrates Rivers contributed to the decline of the Mesopotamian civilization some 6 000 years ago. The estimated global gross area of irrigated land is 270 million ha. About 20 to 30 million ha are severely affected by salinity while an additional 60 to 80 million ha are affected to some degree. Waterlogged soil, which aggravates the problem of salinity, is usually caused by overwatering and a lack of proper drainage systems. Runoff from agricultural areas fertilized with manure and chemicals pollutes watercourses and groundwater by increasing levels of nutrients.

The present level of water pollution warrants that steps be taken to control further contamination of water resources. More serious action needs to be taken in water resource management, waste water treatment and the provision of safe public water supplies. In developed and developing countries there should be controls and regulations regarding the treatment and recycling of industrial effluents, while efforts must be made to replace harmful products and ban dangerous pesticides. There is compelling evidence that at least 20 to 30 percent of the water currently used in households and industries can be saved by adopting appropriate regulatory and policy instruments (tariffs, quotas, groundwater extraction charges). The twin benefits of clean water and reduced demand can be obtained if the recycling or reuse of water is encouraged in industries through pollution control legislation and economic incentives (water tariffs based on economic costs, effluent charges and low-interest loans for effluent and sewage treatment plants). Similar savings may be possible in irrigated agriculture by investments in canal lining, by encouraging less water-intensive crops (through relative output prices) and by raising irrigation rates.

Industry requires large amounts of water, but most of it is recycled back into the water system. The major problem is that much of this water is returned polluted with wastes, chemicals and heavy metals. Over 85 percent of total withdrawals by industry are recycled as waste water.

Domestic water demand is moderate in comparison with agriculture and industry but its quality requirements are high. Domestic and municipal water uses include drinking, washing, food preparation and sanitation.

WATER AND HEALTH

Two of the most troubling domestic water supply issues for policy-makers are access and health. Nearly one billion people in the world are without clean drinking-water. Providing easier access to safe drinking-water significantly improves health conditions. Personal hygiene increases when water availability rises above 50 litres per day (which generally means that it must be delivered to the house or yard). An estimated 1.7 billion persons contend with inadequate sanitation facilities. The lack of sewage collection and treatment is a major source of surface and groundwater pollution. Health officials identify five categories of disease related to water:

(i) water-borne diseases (typhoid, cholera, dysentery, gastroenteritis and infectious hepatitis);

(ii) water-washed infections of the skin and eyes (trachoma, scabies, yaws, leprosy, conjunctivitis and ulcers);

(iii) water-based diseases (schistosomiasis and guinea-worm);

(iv) diseases from water-related insect vectors such as mosquitoes and blackflies; and

(v) infections caused by defective sanitation (hookworm).

The World Bank's World Development Report 1992 estimates that providing access to safe water and adequate sanitation could result in two million fewer deaths from diarrhoea among young children and 200 million fewer episodes of diarrhoeal illnesses each year.

Water as a Strategic Resource

Water, even when plentiful, is frequently drawn into the realm of politics. Domestic laws and well-established customs can help resolve water-related disputes at national and village levels but international law has not developed fast enough to deal with the growing number of water-related conflicts between many countries and regions. In 1989, Egypt's then Minister of State for Foreign Affairs, Boutros-Ghali, declared: "The national security of Egypt is in the hands of the eight other African countries in the Nile basin."14 As Postel notes, Boutros-Ghali highlights the importance of water to Egypt's economy as well as the advantage upstream countries have over downstream neighbours.

The increasing value of water, concern about water quality and quantity, and problems of access and denial have given rise to the concept of resource geopolitics or "hydropolitics". In this context, water joins petroleum and certain minerals as a strategic resource. Its increasing scarcity and value will only intensify the prevalence of water politics and relevant international conflicts.

Several countries depend heavily on river flows from other countries. Botswana, Bulgaria, Cambodia, the Congo, Egypt, the Gambia, Hungary, Luxembourg, Mauritania, the Netherlands, Romania, the Sudan and the Syrian Arab Republic all receive over 75 percent of their available water supplies from the river flows of upstream neighbours. More than 40 percent of the world's population lives in river basins that are shared by more than one country.

Along with land and energy sources, water has been the focus of disputes and, in extreme cases, even wars. The division of the Indus waters and its tributaries among India and Pakistan provided a salutary warning example. War was only just avoided in the early years of independence by a binding agreement, backed by massive international aid, to build two huge water storage dams and a system of canals. Water could then be channelled to the areas of Pakistan that were deprived of water when some of the Indus tributaries were diverted into Indian territory.

The costs to all parties of this settlement were high but certainly less than the human and financial costs of a conflict. Many other international rivers, including the Nile, Euphrates, Ganges and Mekong, are prospective risk points for disputes. The future of the Jordan waters is already an integral component of regional peace talks and illustrates how complicated hydropolitics can be. The fact that groundwater resources are also involved in the talks adds another dimension of difficulty.

THE INTERNATIONAL CONFERENCE ON WATER AND THE ENVIRONMENT: DEVELOPMENT ISSUES FOR THE 21ST CENTURY

The International Conference on Water and the Environment (ICWE) was held in Dublin, Ireland, from 26 to 31 January 1992. The conference provided the major input on freshwater problems for UNCED, convened in Rio de Janeiro, Brazil, June 1992. The ICWE was attended by 500 participants from 114 countries, 38 NGOs, 14 intergovernmental organizations and 28 UN bodies and agencies.

The major work of the ICWE was undertaken by six working groups which addressed:

- integrated water resources development and management;
- water resources assessment and impacts of climate change on water resources;

- protection of water resources, water quality and aquatic ecosystems:
- water and sustainable urban development and drinking-water supply and sanitation;
- water for sustainable food production and rural development and drinking-water supply and sanitation;
- mechanisms for implementation and coordination at global, national, regional and local levels.

The two main outputs of the conference are the Dublin Statement and Report of the Conference, which set out recommendations for action based on four guiding principles. First, the effective management of water resources demands a holistic approach linking social and economic development with the protection of natural ecosystems, including land and water linkages across catchment areas or groundwater aquifers; second, water development and management should be based on a participatory approach that involves users, planners and policy-makers at all levels; third, women play a central part in the provision, management and safeguarding of water; and, finally, water has an economic value in all its competing uses and should be recognized as an economic good.

The Water Sector and Natural Resource Policy

In January 1992, the ICWE concluded that scarcity and misuse of freshwater pose a serious and growing threat to sustainable development and protection of the environment.15 The conference emphasized that human health and welfare, food security, economic development and ecosystems are all at risk, unless water and land resources are managed more effectively in the future.

To address water problems at local, national and international levels, the ICWE recommended a range of development strategies and policies based on four principles. While the conference participants readily agreed on the wording of the first three principles, the fourth provoked a long and contentious debate. Principle 4 declares that water has an economic value in all its competing uses and should be recognized as an economic good.

For many, it is difficult to reconcile the concept of water as an economic good with the traditional idea of water as a basic necessity and human right. Older elementary economic textbooks explain this conceptual puzzle-why diamonds, which have so little utility, are expensive while freshwater, which is so essential to life, is cheap. More recent texts leave water out of these vignettes. Like fresh air, water was once considered a classic free good; now that it is growing scarce, while not yet expensive, it is at least acknowledged to be valuable.

Scarcity is one of the most important issues in considering the various socio-economic tradeoffs in allocating water among different users. Allocation policies and decisions determine who will have access to water and under what conditions, and what impact this will have on society and the economy.

The cheapness of water is often more apparent than real. It is a free good not because water provision is without cost-obviously this is far from true-but because governments have chosen to charge less than full costs for water services for one or more reasons. These subsidies are now coming under scrutiny. The ICWE's final report acknowledges that failure in the past to recognize water's economic value and the real cost of service provision has led to wasteful and environmentally damaging uses. Moreover, the conference report states that managing water as an economic good is an important way of achieving efficient and equitable use, as well as encouraging the conservation and protection of scarce water resources.

Water may be considered a "free" good in the form of rain, but when this free good is captured and delivered to customers by canal, pipe or other means, it becomes a water service. There is generally much less resistance to water service fees than there is to water charges.

It is in this context that the ICWE and UNCED called for a new approach to the assessment, development and management of freshwater resources. The proposed approach involves the management of freshwater as a finite and vulnerable resource and the integration of sectoral water plans and programmes within the framework of national economic and social policy.

FRAGMENTED PLANNING AND WATER RESOURCES IN SOUTHERN INDIA

The World Bank's water resources management policy paper presents several examples from southern India to illustrate the kinds of problem caused by fragmented decision-making. The Chittur River's highly variable flows have traditionally been diverted at many points into small reservoirs to irrigate the main rice crop. The diversion channels are large enough to accommodate flood flows following the monsoon rains. Thus, when a storage dam was constructed, the uppermost channel was able to absorb virtually all the regulated flow. The upper tanks now tend to remain full throughout the year, concentrating benefits and adding to evaporation losses. The more extensive lower areas have reverted to uncertain rain-fed cultivation, and total agricultural value added has decreased. Construction of the storage dam without adequate consideration of downstream users or the existing storage capacity of the basin is one example of how individual project development in isolation can cause significant economic losses.

The construction of the Sathanur Dam on the Ponnani River in Tamil Nadu to serve a left bank command area deprived productive delta areas of irrigation water. While the rights of downstream irrigators are recognized in the dam operating rules, most of the regulated flow is diverted upstream; water losses have greatly increased in the wide sandy bed and no surface water has reached the sea for 20 or more years. Continued spills in about 50 percent of all years were used to justify the subsequent construction of the right bank command, further aggravating shortages in the delta and leading to continual conflicts between the two Sathanur commands. Meanwhile, additional storage dams on upstream tributaries are adding to evaporation losses in what was already a fully developed basin.

Irrigation in the productive delta has declined further and the Sathanur commands in turn are suffering. The high-value crops that were once grown on the main river are being replaced by cultivation on less productive lands, served by tributaries that are more variable than the main river.

The Amaravati River, a tributary of the Cauvery, is the most disputed major river in India. In the absence of a Cauvery agreement, Karnataka (the upstream riparian state) has steadily developed large irrigation schemes, depriving the delta (Tamil Nadu's rice bowl) of its accustomed supplies. Meanwhile, Tamil Nadu has been developing the Amaravati. As at Sathanur, water releases are made from the Amaravati Dam for the traditional areas, but these are far downstream and the substitution of regulated flood flows has encouraged the development of private pumps along the river bank. Even though the new electric connections have now been banned, little can be done to control illegal connections or diesel pumps and, consequently, little water now reaches the lowest commands, let alone the Cauvery. Meanwhile, new storage dams are being constructed on tributaries both in Kerala and Tamil Nadu, further depriving not only the old lands but also the new lands and the pump areas.

A more integrated and broader approach to water sector polices and issues is important because of water's special nature as a unitary resource. Rainwater, rivers, lakes, groundwater and polluted water are all part of the same resource, which means global, national, regional and local actions are highly interdependent.18 Water use in one part of the system alters the resource base and affects water users in other parts.

Dams built in one country frequently reduce river flows to downstream countries for years afterwards, thereby affecting hydroelectric and irrigation capacity. When a city overpumps a groundwater supply, streamflows may be reduced in surrounding areas; when it contaminates its surface water, it can pollute groundwater supplies as well. Certain human actions at local levels may contribute to climate change, with long-term implications for the hydrological system worldwide.

Water policies, laws, projects, regulations and administrative actions often overlook these linkages. Governments generally tend to organize and administer water sector activities separately: one department is in charge of irrigation; another oversees water supply and sanitation; a third manages hydropower activities; a fourth supervises transportation; a fifth controls water quality; a sixth directs environmental policy; and so forth.

These fragmented bureaucracies make uncoordinated decisions, reflecting individual agency responsibilities that are independent of each other. Too often, government planners develop the same water source within an interdependent system for different and competing uses. This project-by-project, department-by-department and region-by-region approach is no longer adequate for addressing water issues.

To help resolve the growing number of water resource issues, policy-makers are increasingly being called on to review and explain the conditions, problems and progress in the overall water sector.

This integrated approach requires water managers to understand not only the water cycle (including rainfall, distribution, ecosystem interactions and natural environment and land-use changes), but also the diverse intersectoral development needs for water resources.

The next section further explores this important concept of linking the water sector with the national economy and provides a conceptual basis for understanding the role of economic policy-making.

WATER RESOURCES: ECONOMICS AND POLICY

In early civilizations, water played a relatively simple role. It was needed for transportation and drinking and it provided a fishing and hunting source. Overtime, sedentary agricultural societies evolved and water use became more important.

Families began settling near springs, lakes and rivers to supply livestock and crops with water, gradually developing technologies to divert water for irrigation and domestic purposes. Babylonian, Egyptian, Hittite, Greek, Etruscan, Roman, Chinese, Mayan, Incan and other empires constructed water delivery systems such as long aqueducts to carry water to large cities. In fact, until the middle of the twentieth century, most societies were able to meet their growing water needs by capturing reliable and relatively inexpensive sources.

When water is plentiful relative to demand, water policies, rules and laws tend to be simple and only casually enforced. As populations grow and economies expand, water sectors evolve from an "expansionary" phase to a "mature" phase. At a certain point during the expansionary phase, the financial and environmental costs of developing new water supplies begin to exceed the economic benefits in the least productive (marginal) uses of existing supplies. The real location of existing supplies, rather than the capture of unclaimed supplies, then becomes the least costly method to maximize benefits.

A water sector in the "mature" phase is characterized by rising marginal costs of providing water and increasing interdependencies among users. In this phase, conflicts over scarcities and external costs arise. (External costs result when one user interferes with another's supply, e.g. when an upstream user pollutes a river and raises costs for downstream users.) These conflicts eventually become so complex that elaborate management systems are needed to resolve disputes and allocate water among different users and economic sectors.

Developing effective water sector policies is troublesome for a number of reasons. First, water has unique physical properties, complex economic characteristics and important cultural features that distinguish it from all other resources. Second, water resource management is administratively complicated because it involves legal, environmental, technological, economic and political considerations. In most societies, political considerations dominate decisions on water resource use. Nonetheless, most policy options are framed and discussed in economic terms. For example, water resource management depends on the government's ability to establish an appropriate legal, regulatory and administrative framework. In fact, markets are based on a system of enforceable private property rights. Private water markets require secure and transferable property rights, including the right to exclude other users.

This section attempts to provide a conceptual basis for understanding water policy interventions while examining the circumstances under which water policies work or fail. It comprises three parts: the first examines the relationship between the water sector and the overall economy; the second explains the social, physical and economic nature of water; the third assesses the advantages and disadvantages of broad alternative approaches to public water policy and also reviews policy issues related to the economic organization of water resource management.

Linking the Water Sector with the National Economy

Economic policy-makers tend to confront policy issues one at a time, stating policy objectives in single dimensional terms. This approach presents difficulties because a policy aimed at achieving a single objective usually has unintended and unrecognized consequences. Water managers and policy-makers need to assess the entire range of government interventions to understand fully the economic, social and environmental impacts on a given sector, region or group of people.

Improving water resource management requires recognizing how the overall water sector is linked to the national economy. Equally important is understanding how alternative economic policy instruments influence water use across different economic sectors as well as between local, regional and national levels and among households, farms and firms. For too long, many water managers have failed to recognize the connection between macroeconomic policies and their impact on, for example, technical areas such as irrigation.

Macroeconomic policies and sectoral policies that are not aimed specifically at the water sector can have a strategic impact on resource allocation and aggregate demand in the economy. A country's overall development strategy and use of macroeconomic policies-including fiscal, monetary and trade policies-directly and indirectly affect demand and investment in water-related activities. The most obvious example is government expenditures (fiscal policy) on irrigation, flood control or dams.

A less apparent example is trade and exchange rate policy aimed at promoting exports and earning more foreign exchange. For example, as a result of currency depreciation, exports of high-value, water-consuming crops may increase. If additional policy changes reduce export taxes, farmers are provided with an even greater incentive to invest in export crops as well as in the necessary irrigation. National development strategies can directly influence water allocation and use in other ways. In the case of a food self-sufficiency strategy, the government may subsidize water-intensive inputs to encourage farmers to produce more rice. By providing financial incentives for rice producers, the government is influencing the demand for water and private irrigation investment through price policies.

Apart from the direct effects on water use resulting from such price policies, the increased demand for irrigation water also has intersectoral, intrasectoral, distributional and environmental implications. The agricultural sector is provided with an economic advantage in access to water vis-à-vis the industrial sector (intersectoral); water used for rice gains an economic advantage over water used for other crops (intrasectoral); rice producers with more land and access to water gain over those with less land and water (distributional); and increased pesticide and fertilizer use are likely to affect water quality (environmental).

Sectoral policies affect water use and allocation in non-agricultural sectors in a variety of ways. For example, in the western United States, 70 to 80 percent of the region's water yield results from snowmelt from the high-elevation forests, many of which are under public jurisdiction. Water yields are significantly affected by timber harvest policies on these lands. Rangeland management policies on lower elevations also alter vegetation conditions and thus affect the rate of evapotranspiration, in turn affecting streamflow and groundwater recharge. In such cases, it is important for downstream city water managers to recognize, understand and become involved in the decisions of other sectors such as livestock and forestry.

Economic policies and water use in the syrian arab republic

After struggling throughout the 1980s, the Syrian economy has performed well over the past few years. The end of a two-year drought allowed agriculture and agro-industries to recover in 1991. During the drought, the government was forced to import large quantities of wheat and barley, thereby draining foreign currency reserves. In addition, the lower water levels meant a reduction in hydropower generation, increasing the need for thermal power and, in turn, lowering crude oil exports.

Two of the Syrian Arab Republic's major national development objectives are: achieving food self-sufficiency to reduce dependency on imports; and expanding agricultural exports to earn more foreign exchange. To support these objectives, the government has invested 60 to 70 percent of the entire agricultural budget in irrigation over the past ten years.

Several factors explain this special attention for irrigation development. The irrigated area comprises only 15 percent of the cultivated land yet produces over 50 percent of the total value of agricultural production. A large part of wheat production as well as all major industrial crops, including cotton, tobacco and sugar beet, are produced on irrigated farms. Production on the remaining rain-fed area, representing 85 percent of the total area, varies greatly from year to year. At present, agriculture accounts for about 85 percent of the country's water consumption, but competition is increasing. During the 1980s, industrial water demand increased by nearly 900 percent. Current projections suggest that water requirements will be two to three times greater by 2010.

The government's effort to promote food self-sufficiency has produced a second generation of water-related problems. To encourage growth in agricultural production and enhance rural incomes, interest rates, seeds, fertilizers, pesticides, transport and energy prices are subsidized. The government also establishes purchase prices and buys industrial crops, major cereals and feedgrains; e.g. the 1992 domestic wheat price was almost twice the international price.

These policies are contributing to the proliferation of wells in the Syrian Arab Republic. Digging wells to pump groundwater accounts for 80 percent of the newly irrigated land since 1987. With irrigation, farmers obtain higher yields, more stable production and greater profit. Since water is free, the only investment expense required is the well and the pumping gear-a one-time fixed cost. Farmers obtain subsidized credit to purchase subsidized fuel for operating imported pumps purchased with overvalued currency (an implicit subsidy). With these economic opportunities, most farmers want to dig wells or pump surface water.

Other current economic pressures are also influencing farmers' decisions to dig wells and expand irrigation. For example, as incomes in urban areas increase, consumers are demanding more fruit and vegetables. At the same time, recent changes in trade and exchange rate policies are making Syrian agricultural products more competitive in regional markets. Farmers who inititally planned only on supplementary irrigation for winter wheat are finding summer vegetables and irrigated fruit production increasingly profitable.

With the continuing importance of structural adjustment and stabilization programmes, many developing countries are implementing fundamental changes in macroeconomic and sectoral polices. Typical adjustment programmes call for a greater reliance on markets, more open trade, fiscal austerity and a phasing out of producer and consumer subsidies (input and product markets). Budget-reducing measures imply increased competition between and within sectors for funding new water projects. In these situations, the overall economic, social and environmental implications of choices must be carefully addressed. For example, when governments must choose between financing either irrigation projects or hydroelectric power projects, there is an additional social opportunity cost of the irrigation water in countries that are dependent on imported energy sources.

At the same time, when water scarcity keeps some farmers on uneconomical lands such as steep watersheds, the country suffers twice: once in terms of reduced production compared with what would be possible with irrigation; and again in terms of erosion and resource depletion, with erosion possibly shortening the life of existing waterworks.

In most countries, pressure has increased not only to modify investment allocations but also to recognize and accommodate new demands for water. The direct implications for water managers include fewer capital investments in new water projects, the elimination of irrigation subsidies, increased efforts to recover its cost and more emphasis on demand management to improve the efficiency of existing supplies.

The Social, Physical and Economic Nature of Water

Policy-makers throughout the world treat water as more than a simple economic commodity. Because water is essential to life, they often reject competitive market allocation mechanisms. Many societies believe that water has special cultural, religious and social values. Boulding observed that "the sacredness of water as a symbol of ritual purity exempts it somewhat from the dirty rationality of the market". In many cultures, goals other than economic efficiency play an unusually large role in selecting water management institutions. Some religions, such as Islam, even prohibit water allocation by market forces.

The international community recognizes that access to water is a basic human right. The ICWE asserted that "...it is vital to recognize first the basic right of all human beings to have access to clean water and sanitation at an affordable price".

The connection between water and human life is most dramatic in arid regions, where crop irrigation is essential to food production. In Egypt, little food can be grown without the help of the Nile for irrigation. However, the focus on water's special status tends to obscure the fact that, in most societies, only a tiny fraction of water consumption is actually for drinking and preserving life. In fact, a large portion of urban water is used for convenience and comfort. In the arid western United States, per caput water withdrawal by households frequently exceeds 400 litres per day, about half of which is used to irrigate lawns and gardens. Most of the remainder is for flushing toilets, bathing and washing cars.

Another important influence on water resource policy is societies' partiality for technical solutions. In most countries, water management is typically relegated to the engineering domain. Indeed, most water managers are engineers, who are trained to solve technical problems. As inadequate public policies are increasingly blamed for water-related problems, a strong case is emerging for emphasizing human behaviour as an additional component of water systems.

Physical Attributes of Water

Water has two additional features that further complicate management efforts: bulkiness and mobility. The value per unit of weight tends to be relatively low (placing water among the commodities that are termed "bulky"). Unlike petroleum, the costs of transporting and storing water are generally high relative to its economic value at the point of use. In crop irrigation, the water applied may yield additional economic values of less than $0.04 per tonne of water. Water is also difficult to identify and measure because it flows, evaporates, seeps and transpires. This evasive nature means that exclusive property rights, which are the basis of a market economy, are hard to establish and enforce.

Many water management problems are site-specific and so elude uniform policy treatment. While water consumption and quality requirements are tied to local populations and development levels, local water availability usually changes with climatic variations throughout the year and over longer cyclical swings. These supplies may be highly variable and unpredictable in time, space and quality. In regions throughout India, for instance, most rainfall is concentrated during a three-month period and there are large year-to-year variations. In addition, forecasts of significant global climate change-attributable to both natural and human causes-raise concerns about longer-term supply trends.

Water projects that attempt to compensate for extreme seasonal variations such as floods and droughts frequently require enormous investments. The economies of size are so large in these cases that unit costs continue to exceed the range of existing demands. This is a classical "natural monopoly" situation in which a single supplying entity is the most economically efficient organizational arrangement. On the other hand, most economies of size for pumping groundwater are achieved at relatively small outputs and multiple suppliers can therefore operate efficiently. However, aquifers are usually hydraulically linked with rivers or streams-part of a river's volume may come from underground flows and rivers may replenish groundwater stocks. This hydraulic linkage is affected when an aquifer is heavily pumped. A lowered groundwater table may draw water from a connected stream, reducing its flow to surface water users.

Climate change, water resources and agriculture

To date, research has not been able to provide clear conclusions about the prospective impacts of climate change and global warning. Among the potential impacts of climate change is its effect on the hydrological cycle and water management systems. For instance, an increase in floods and droughts will increase the frequency and severity of disasters. Relatively small changes can cause severe water resource problems, especially in semi-arid regions and humid areas where demand or pollution has led to water scarcity.

The statement adopted by the Second World Climate Conference, held in Geneva in 1990, concluded that the design of many costly structures to store and convey water, from large dams to small drainage facilities, is based on analyses of past records of climatic and hydrological parameters. Some of these structures are designed to last from 50 to 100 years or even longer. Records of past climate and hydrological conditions may no longer be a reliable guide for the future. The possible effects of climate change should be considered in the design and management of water resource systems.

Data systems and research must be strengthened to predict water resource impacts, detect hydrological changes and improve hydrological parameterization in global climate models.

Agricultural impacts could be significant but researchers are uncertain whether global agricultural potential will increase or decrease. Increases in drought risk are potentially the most serious effect of climate change on agriculture. Disease and pest patterns, raised sea levels and storm surges are additional problems. It also appears that many areas will have increased precipitation, soil moisture and water storage, thus altering patterns of agricultural ecosystems and other water uses.

Aquifer management is often complicated by the aggregate impact of the actions of many individuals. Even though each individual may have a negligible impact when taken alone, the sum total can be of major importance. One example is the rapid spread of tube well irrigation in South Asia. One tube well has little effect on the total water supply, but thousands of tube wells can quickly deplete an aquifer. Establishing effective policies to regulate these many small, scattered decision-makers is exceedingly difficult.

ECONOMIC ATTRIBUTES OF WATER USE

Water provides four types of important economic benefits: commodity benefits; waste assimilation benefits; aesthetic and recreational benefits; and fish and wildlife habitats. Individuals derive commodity benefits from water by using it for drinking, cooking and sanitation. Farms, businesses and industries obtain commodity benefits by using water in productive activities. These commodity benefits represent private good uses of water which are rivals in consumption (e.g. one person's or industry's water use precludes or prevents its use by others). Government policies and regulations that concentrate on improving market access and competition are important means for improving the productive and allocative efficiency of the commodity uses of water.

The second and increasingly important economic benefit of water is waste disposal. Water bodies have a significant, but ultimately limited, assimilative capacity, meaning that they can process, dilute and carry away wastes.

Recreation and aesthetic benefits and fish and wildlife habitats were once regarded as luxury goods outside the concern of governments. Today, these two types of benefit are gaining increased attention. In developed countries, more and more people are focusing their recreational activities around lakes, rivers and seas. In developing nations, as incomes and leisure time grow, water-based recreation is becoming increasingly popular and an adequate supply of good-quality water helps provide a basis for attracting the tourist trade. Examples are cruises on the Nile in Egypt and visits to the Iguazú Falls on the Brazil-Argentina border. Likewise, information and knowledge about how humans have an impact on ecosystems have raised concern about the fish and wildlife benefits provided by water. Fish and wildlife habitats are related to both commodity and recreational uses.

Waste assimilation and recreational and aesthetic values are closer to being public goods than private goods. Public goods are non-rivals in consumption-one person's use does not preclude use by others. For example, the enjoyment of an attractive water body does not deny similar enjoyment to others. Non-rival goods require large amounts of resources to exclude unentitled consumers from using the good. Exclusion costs are frequently very high for water services such as flood control projects and navigation systems. Goods and services that are non-rivals in consumption are normally better suited to public sector interventions, including ownership, provision and regulation.

Aquifer Overdraft

An aquifer is a geological formation, actually or potentially containing water in its pores and voids. Aquifers consist of the porous rock or soil media (sand, gravel or rock materials) within which water is collected and through which it flows. Moisture from rain or snow that escapes evaporation collects in streams as surface water or seeps into the ground. Soil water not taken up by plants seeps downwards until it reaches the water-saturated zone. Water in aquifers is called groundwater. Groundwater deposits are economical to use for human purposes if they are close to the surface (and thereby inexpensive to pump) and are of good quality.

Aquifers vary greatly in their nature and extent. The quantity, quality and ease of extraction can be determined accurately only after extensive exploration. Underground geology varies widely and is expensive to map. Aquifers may be very thin or hundreds of metres thick; some are local in character, while others extend for hundreds of kilometres. The Ogallala-High Plains Aquifer in the central-western United

States underlies more than 10 million ha over six states. Relative to surface water, groundwater moves very slowly-in some cases only a few metres per year. While aquifers may have accumulated over thousands of years, modern pumping devices can easily exhaust them more rapidly than the natural recharge rate. It is also possible to divert surface water to recharge an existing aquifer artificially and make it available for future use.

Aquifer status reports from many parts of the world suggest that all is not well with our groundwater resources. Symptoms of management problems begin with pumping rates that exceed the natural recharge. Primary symptoms are: an exceedingly rapid exhaustion of groundwater stocks and the consequent increase in pumping costs; the intrusion of poorer-quality water into the deposit being exploited; salt water intrusion from rapid pumping near seacoasts; and mineralized deposits interspersed with better-quality water.

Subsidence of overlying lands is another adverse impact of aquifer overexploitation. As water is withdrawn, the soil and rock particles comprising the aquifer are compressed into a smaller volume and, consequently, crack the earth's surface. This results in damage to buildings, roads, railroads, etc. Another consequence of overpumping may be the interruption of flows in neighbouring wetlands and streams; deprived of their water source, they are reduced in size or may dry up altogether. Other adverse effects from overpumping result when residential or farmers' wells dry up because of the presence of larger and deeper wells.

From a broad perspective, aquifer exploitation can bring about either or both of two types of social dilemma. First, overdraft is an example of a class of resource problems, usually called "common pool" problems. A common pool resource can be defined by two characteristics. The first is subtractibility (meaning that a unit of resource withdrawn by one individual is not available to another individual user). The second is the high cost of excluding potential beneficiaries from exploiting the resource. Fugitive or mobile resources, such as water, petroleum or migratory fish and wildlife, are typical examples of resources with high exclusion costs. Common pool problems or dilemmas arise when individually rational resource use leads to a non-optimal result from the perspective of the users as a group. Three conditions are necessary to produce a common pool resource dilemma: first, large numbers of users withdraw the resource; second, the actions and characteristics of the individual users and the extraction technology bring about suboptimal outcomes from the group's viewpoint; third, there must be an institutionally feasible strategy for collective resource management that is more efficient than the current situation.

The roots of the problems associated with common pools are found in the inadequate economic and institutional framework within which the resource is exploited. Common pool resources have been typically utilized in an "open access" framework, within which resources are used according to a rule of capture. When no one owns the resource, users have no incentive to conserve for the future and the self-interest of individual users leads them to overexploitation. The characteristics of the economic institutions governing their use is the fundamental issue in managing common pool resources.

The second type of social dilemma associated with groundwater exploitation is the imposition of external costs or externalities. In the presence of significant externalities, the calculation of costs and benefits by exploiters does not yield a collectively optimal rate of exploitation.

Economic Organization of the Water Sector: Markets or Governments?

Most countries rely on a mix of market policies and direct government interventions to manage water resources. Each system has its own advantages and disadvantages.

A competitive market has the potential to allocate resources (water supplies) efficiently among competing demands. Producers and consumers acting in their own self-interest reach the price at which available supplies are allocated. Private producers, guided by prospective profit, seek to buy inputs as cheaply as possible, combine them in the most efficient form and create products that have the highest value relative to cost.

Consumers' incomes, tastes and preferences influence expenditure patterns, which encourage firms to produce the commodities people are willing and able to buy. Prices are forced upwards for the commodities most desired, and producers allocate resources in the direction of the greatest potential profits. The firms producing desired goods most efficiently are rewarded by profit while the unsuccessful are eliminated, so production occurs at the least cost. However, the needs of potential consumers with limited incomes may either not be met at all or be met only partially.

While the private market has the potential to produce the maximum private-valued bundle of goods and services, the public sector also plays an important role. Public actions incorporate a broader range of social goals than the private sector. The public sector can ameliorate income inequalities, promote development in disadvantaged regions, regulate private activities that harm the environment and control other undesirable effects of a private, profit-oriented monopoly.

Market Failures

If water as a commodity, or the economic system in which water is used, meets the preconditions for a market system, government interventions can be minimized. In competitive markets, government's primary role is to emphasize "incentive structures" and to establish "rules". Some of the most important rules are the laws governing the establishment of property rights and the enforcement of contracts.

Market economies experience shortcomings called market failures.26 Market failures occur when incentives offered to individuals or firms encourage behaviour that does not meet efficiency criteria or, more generally, because efficiency or economic criteria fail to satisfy national social welfare objectives. In these cases, the public sector may intervene to influence water provision and allocation. Market failures affecting water resources include externalities, public goods and natural monopolies. In other cases, even efficient markets may not meet societies' equity criteria so public intervention is necessary to compensate for distributional inequity.

Externalities are inherent in water sector activities. An example is the detrimental effect of saline return water flows (caused by irrigation) on downstream water users. Another example is the waterlogging of downslope lands through inefficient irrigation practices. Most irrigators do not normally consider the external costs they impose on others, so governments attempt to protect affected individuals through regulations, taxes, subsidies, fees or technical standards. For instance, irrigation practices can be regulated by setting and enforcing standards to control salinity and waterlogging.

In recent years, the "polluter pays" principle has attracted increased attention in industrialized countries (and to a lesser extent in developing countries). This principle requires producers to pay the "full" cost of their production process, including externalities such as polluting water.

Water storage projects and flood control programmes represent examples of public goods. The market does not adequately supply public goods because private entrepreneurs cannot easily exclude non-paying beneficiaries and capture a return on investment. For example, it is not possible to exclude people living along a river from the benefits of a flood protection plan on that river.

A firm that experiences decreasing costs throughout its range of production is easily able to dominate the entire market and become a natural monopoly (a common situation in the water sector). Decreasing costs imply increasing returns; thus, the first firm to begin production can always underprice new entrants. Urban water supply systems, hydropower plants and canal irrigation projects are subject to this type of market failure. Unregulated monopolies can restrain production and charge excessive prices; they also have little incentive to innovate. A water supplier acting as a natural monopoly has the power to impose exorbitant costs-even economic ruin-on its customers.

Public regulation or public ownership can mitigate the undesirable effects of a private, profit-oriented monopoly. When increasing returns exist, the lowest-cost production is that of a single producer. Society is likely to benefit by regulating or owning the monopoly rather than by encouraging competitive suppliers. More than one competitive supplier would present much higher distribution costs.

While free competition is viewed as the most efficient system for allocating resources, potential market imperfections can accentuate income disparities. Societies' public welfare goals often incorporate a broad range of social objectives. Primary among these is ameliorating income inequalities between members of the society and sometimes among political subdivisions or regions. In these situations, the government may direct investment and subsidies towards specific regions or groups. Water projects provide important investment strategies both for human welfare (drinking-water and food supplies) and for infrastructure to support economic development.

Government Failures

Even in the event of market failures, public sector interventions or non-market approaches may not lead to the socially optimum solution. In many cases, non-market responses to market failures lead to less than optimal outcomes. In particular, some government agency performance incentives result in a divergence from socially preferable outcomes (both in terms of allocative efficiency and distributional equity criteria). The problem areas relevant to water sector services are:

- *"Products" are hard to define.* The outputs of non-market activities are difficult to define in practice and difficult to measure independently of the inputs that produced them. Flood control or amenity benefits of water storage reservoirs are examples of water system outputs that are hard to measure.
- *Private goals of public agents.* The internal goals, or "internalities", of a public water agency as well as the agency's public aims provide the motivations, rewards and penalties for individual performance. Examples of counterproductive internal goals include budget maximization, expensive and inappropriate "technical-

fix" solutions and the outright non-performance of duties. In addition, agencies may adopt high-tech solutions, or "technical quality", as goals in themselves. For example, they may recommend sprinkler or drip irrigation systems when other less expensive but reliable methods are more economical. Finally, irrigation agency personnel may be persuaded, by gifts or other inducements, to violate operating rules for a favoured few.27

- *Spillovers from public action.* Public sector projects can also be a major source of externalities. Salinity and waterlogging of downslope lands can occur just as easily from inappropriately managed public irrigation projects as from private irrigators.
- *Inequitable distribution of power.* Public sector responsibilities, however noble their intent, may not be scrupulously or competently exercised. Yet the monopoly control of water supplies by public agencies provides certain groups or individuals with so much power over the economic welfare of water users that procedures to protect those of limited influence should be of prime importance.

Economic Structure and Irrigation

For many years, the economic systems in a number of developing countries discriminated against agriculture through policies such as high levels of protection for domestic manufacturing sectors, overvalued exchange rates and taxes on agricultural exports. Most developing countries today are at some stage of structural reform, attempting to adjust and transform their economies towards a more liberal economic trade regime-modifying government involvement and increasing market influence.

The developing world's recent record in consolidating macroeconomic stability with solid economic growth is very mixed. Where success is evident, most of the economic transformation has taken place at the macro level and much remains to be done to effect the consequent adjustments at the micro level, at the level of water users in other words.

Even with widespread acceptance of the need for macroeconomic price policy reforms for all other sectors since the early 1980s, the dominant supporting actions for agriculture have been non-price policies. For non-agricultural sectors, the new policy mix includes minimizing state involvement in the pricing and marketing of inputs and outputs, privatization and limiting government borrowing.

Despite the irrigation sector's often being sheltered or even benefiting from the effects of these economic policy reforms, government subsidy cuts are inevitably affecting the scope and efficiency of agricultural support services. In most countries, there is a pressing need to discuss how various policy options, including both public interventions and market-oriented, private sector activities, may assist the irrigation sector in the process of economic reform.

2

Supply and Demand of Water

To help select the most appropriate policy option or programme alternative, policy analysts divide the water sector into supply-side and demand-side components. The supply-side approach is structure-oriented; investments in water projects are combined with engineering and technical expertise to capture, store and deliver water and to make systems operate effectively. The supply side focuses on providing water and related services.

For most of the twentieth century, policy-makers have focused their attention on the supply side. Economists have evaluated public water supply and policy options through benefit-cost analysis (BCA). The main purpose of BCA is to assure that scarce resources (such as labour, capital, natural resources and management) are all employed to their best advantage. BCA attempts to quantify the advantages and disadvantages to society of alternative policies or actions in terms of a common monetary unit.

With new water-related problems arising in many parts of the world, policy-makers are increasingly emphasizing non-structural approaches to water management. A non-structural approach encompasses demand management, scientific research, education and persuasion to coordinate how humans use water. These demand-side policies attempt to address the human causes of water problems such as water quality degradation, cverexploitation of aquifers and the decreasing availability of water flows to meet non-consumptive water uses (hydroelectric power, pollution assimilation and fish and wildlife habitats). In part, this user-focused approach entails coordinating and influencing people through organizations and institutions. This focus on water users aims to promote least-cost, environmentally sound water planning and takes both demand and supply options into account.

Voluntary associations, government bureaucracies and private businesses are examples of organizations that operate on both the demand (user) side and the supply (delivery) side of water supply systems. On the supply side, large hierarchical organizations usually control the capture, storage, conveyance and distribution of surface water. Demand-side organizations, for example water user associations, are established to represent the interests of irrigators and to introduce and enforce water allocation rules.

This section examines how water institutions and water allocation systems incorporate a user-focused approach to address surface water, groundwater and water quality issues related to agriculture.

INSTITUTIONS AND WATER POLICY

Institutions are defined more broadly than simply government agencies and private organizations. Institutions are "...sets of ordered relationships among people which define their rights, exposure to rights of others, privileges and responsibilities". In this context, institutions set the "rules of the game" within which the economic system operates. For example, the property rights system is considered a water institution because it includes provisions which determine access to water and land. The property rights structure helps define the incentives, disincentives, rules, rights and duties (including informal customs and formal legal systems) that guide human activities and encourage conformist behaviour.30 Thus, property rights are part of an institutional arrangement governing economic activities including water use.

Many demand-side policies attempt to coordinate water use through institutions such as property rights and incentive structures such as prices. Altering the institutional system of permissions, restrictions, incentives and penalties can compel consumers to do what they might not otherwise do. For instance, financial inducements (monetary rewards and penalties) can encourage people to use water in a more socially desirable way.

Two factors significantly influence the form of water institutions in a society: the relative scarcity of water and the transactions costs required to establish and enforce water rights. While scarcity is both supply-and demand-dependent, the human pressures on the demand side are probably the most important. Transactions costs include the resources required to obtain information, negotiate agreements on property rights and police these agreements. Water supply and demand characteristics make transactions costs for water relatively high and the value of water relatively low compared with other resources or commodities.

Many economists are trying to find ways to improve water-use efficiency through improved institutional performance. Research to date suggests that institutions and technological change are altered in response to the same types of incentives. When water is plentiful relative to demand, laws governing water use tend to be simple and enforced only casually. Where water is scarce, more elaborate institutional systems evolve. Higher population and income levels as well as technological advances are prompting many governments to establish formal water-use and water quality management systems.

Establishing an institutional structure for allocating water is a fundamental role of social policy for any nation. The choice of structure is ultimately a compromise between the physical nature of the resource, human reactions to policies and competing social objectives. Not surprisingly, different cultures make tradeoffs based on the relative importance of their particular objectives.

Countries try various means to balance economic efficiency (obtaining the highest value of output from a given resource base) and fairness (assuring equal treatment). Individual freedom, equity, popular participation, local control and orderly conflict resolution are other important objectives which societies must juggle when choosing a structure for water allocation.

Water Allocation Systems

In an "ideal" market-based water allocation system, entitlements (water rights) are well defined, enforced and transferable and they confront users with the full social cost of their actions. This type of market-dependent institutional arrangement requires security, flexibility and certainty.34 Security refers to protection against legal, physical and tenure uncertainties.

The assumption is that users will undertake profitable long-term investments to obtain and use water supplies only if water entitlements are reasonably secure.

A system is flexible if allocations between users, uses, regions and sectors can be changed at a low cost in relation to benefits. Flexibility implies that changes in demand are accommodated easily by reallocating water to higher-valued uses as they emerge. Certainty is also necessary: water-use rules must be easy to discover and to understand.

The three basic types of "water rights" systems are: (i) riparian-only those owning the land in physical contact with a natural watercourse have a right to use it; (ii) prior appropriation-based on beneficial and actual use; and (iii) public administration-a public authority authorizes water distribution and use. Prior appropriation and public administration are the most common systems in use throughout the world.

Property Rights Systems and Surface Water Allocation

Some water allocation systems are relatively decentralized and based on entitlements or rights to specific quantities of water. Examples of these types of system are found in the western United States and southern Australia. In contrast, France uses a more centralized public utility model of water allocation. Chile is the only country with a comprehensive water allocation system that establishes tradeable property rights.

Chile's water law: allows trade between and among economic sectors; protects third party rights; establishes compulsory water user associations and a national water authority to resolve conflicts; and allows for judiciary solutions to those conflicts not resolved by water user organizations or the water authority. Water transfers require authorization at two levels-those of the local water user associations and the national water authority.

While formal tradeable water rights systems have not been established in other developing countries, Rosegrant and Binswanger document the expansion in surface water and groundwater markets. One recent study on surface water trade in Pakistan reported active markets in 70 percent of watercourses.

A "public utility" model for water allocation and pollution control

France's model for allocating water could be termed a "public utility" model. Administration is centred in six river basin committees (RBCs) and six river basin financial agencies (AFBs) which control water abstraction, treatment and delivery. The RBCs are the centre for negotiations and policy-making regarding water management at the basin level. The AFBs base action plans on extensive water data (quality and quantity needs) and are the centre of knowledge and technical expertise for the government and other interested water users. The RBCs collect fees, award grants and loans, develop long-term plans, collect and analyse water data, conduct studies and finance research.

The RBCs approve 20-or 25-year water development plans and, every five years, establish action plans to improve water quality. They also set two fees to be paid by water users: one for water consumption and the other for point source pollution. The fees provide incentives for users and also form a fund to encourage better water use through grants or soft loans. The RBCs include representatives from national, regional and local government administrations as well as individuals from industrial, agricultural and urban interests.

Those withdrawing water outside the authority (excluding small units) must measure and pay for it. Costs depend on the source (surface or ground) and how valuable it is in the specific basin. This approach also employs the "polluter pays" principle. While pollution assimilation is recognized as a legitimate water use, the entity causing pollution must pay for the costs of remediating the pollution and must compensate for any damages. Reports indicate that the system is self-financing and that it has performed well over 25 years.

The French approach provides feasible and apparently effective solutions to the major concerns about public water management: water scarcity, pollution and conjunctive use of ground and surface water.

It does, however, appear to rely more on centralized administrative discretion and less on the preferences and initiatives of private individuals. For example, water charges are set without regard for scarcity values based on bids and offers by users.

The water allocation system in the western United States originated last century and evolved out of the customs of miners and farmers. This system is called the "prior appropriation" doctrine because water entitlements are granted according to the date on which a person applies the water to a beneficial use. The phrase "first in time, first in right" describes the basic principle, as the date of appropriation establishes the order in which users may draw from the water source.

In the western United States system, the individual's property interest in water is limited to the right to divert and use a specified quantity. Private individuals cannot "own" water but have "usufructuary" rights. The state retains ownership and determines which uses are beneficial. Beneficial uses were originally limited to private sector, off-stream purposes in agriculture, households and industry. More recently, in-stream uses for recreation and for fish and wildlife habitats are being recognized.

Water rights are generally tied to a specific parcel of land. However, in most states these water rights can be sold, without loss of priority, to another individual for use on another parcel. Rights are protected by the state from other appropriators and cannot be taken from an individual by the government without just compensation for the foregone economic value.

In the prior appropriation system, irrigation water rights are administered by private non-profit cooperative organizations or by public districts, under the supervision of the state government. Financing requirements and cost recovery are normally based on area served rather than on a strict volumetric pricing system. Early in the history of western irrigation, private capital was the main source of funds. In the twentieth century, private financing has mostly been replaced by federal subsidies. In contrast, municipal and industrial water supplies are typically financed by the users, with a full-cost pricing rule.

In recent years, southern Australian states have begun to allocate water through a system of transferable water entitlements. The Australian water law is based on a non-priority permit system under state control and ownership. Individuals obtain a right to use water through a licence issued by a state agency; this right is usually for a specified type of use on a designated tract of land. In contrast to the United States doctrine of prior appropriation, all users share equally in supply shortfalls and permits expire after a specified time period. Recent studies indicate the need to base permits on the water system's capacity in proportion to the quantities assigned in the original permit.

In southern Australia, the transfer of rights is generally limited to water users in the same watershed, with special conditions imposed to protect supply reliability and to prevent third-party damages. Reports evaluating water market performance highlight two points: first, the system is facilitating reallocation from low-value to higher-value uses; and, second, transferable entitlements should be considered as part of a broad package of decentralized decision-making for the entire water sector.

Prices and Surface Water Allocation

In practice, market forces rarely establish prices for water. Instead, prices are set by publicly owned supply agencies or regulated private utilities. Water prices ("rates" in public utility jargon) have an impact on both efficiency and equity as well as influencing agency revenues. The charging scheme for recovering costs and allocating water is a decision variable for the supplying or regulating agency.

Rate-setting can be evaluated within a multiple objective framework in which allocative efficiency, equity of income distribution and fairness in apportioning costs all play a role in evaluating pricing policies. The secondary criteria of simplicity, administrative feasibility and stability are also taken into account.

The most commonly employed pricing policy for water is a flat rate charge, designed primarily to recover costs. Flat rates are not set according to the volume received, although a proxy for volume usually provides the basis for the charge. In agriculture, the most frequent basis for a water charge or service fee is the area irrigated.

For residential use in the industrialized world, flat rate charges have been based on the number of residents, the number of rooms, the number and type of water-using fixtures or measures of property value.

Flat rates are criticized because they do not include incentives for rationing water in line with willingness to pay. Such schemes are, however, simple to administer and assure the supplier adequate revenue.

The high cost of installing and monitoring meters is suggested as being the main reason for continuing the flat rate approach. This argument is convincing in cases where water is plentiful, supply costs are low and managers doubt the rationing effects of volumetric pricing. In other cases, water managers are turning to volumetric pricing to address water scarcity problems and the high costs of developing new supplies.

Policy-makers who are primarily interested in allocative efficiency (maximizing net social product) as the goal for a pricing scheme advocate marginal cost pricing. The marginal cost represents the incremental cost of supplying a good or service. The marginal cost is a schedule of costs related to quantity and typically rises as further increments are supplied. When water prices are set at the marginal cost, rational consumers demand additional water only as long as willingness to pay (demand) exceeds the incremental costs. In theory, marginal cost pricing yields the most economically efficient allocation.

Price impacts on water use

In Australia, Canada, Israel, the United Kingdom and the United States, studies have demonstrated that water demand drops by 3 to 7 percent when prices charged to households rise by 10 percent. While it is difficult to measure elasticity of demand without metering consumption, some research in developing countries indicates how prices, combined with other policy efforts, affect water use.

China

In Beijing a water quota and high rates for exceeding it led to a 37 percent reduction in industrial water use in the 1980s. During the same period, the industrial sector was able to expand rapidly.

India

A fertilizer plant at Goa reduced water consumption by 50 percent in response to higher water prices. The Goa plant now uses 10.3 m3 to produce 1 tonne of nutrient, paying $0.12 per m3. In contrast, a similar plant at Kanpur pays $0.01 per m3 but uses 24.35 m3 per tonne of nutrient.

Indonesia

In Bogor a water tariff increase ranging from 200 to 300 percent (from $0.15 to $0.42 for the first 30 m3 per month) decreased monthly consumption by around 30 percent for domestic and commercial connections.

Source: R. Bhatia and M. Falkenmark. 1992. Water resource policies and the urban poor: innovative approaches and policy imperatives. Background paper for the ICWE, Dublin, Ireland.

A number of obstacles are encountered in the application of marginal cost pricing. One problem is the variety of definitions of the appropriate marginal cost concept, particularly whether to use a short-run (variable cost) concept or a long-run, full-cost approach. A long debate ensued from the "short-run marginal cost" pricing proposal which emerged from welfare economists' work in the 1930s. For example, Coase strongly objected to setting utility prices at short-run marginal costs, especially where marginal costs are below average costs (thereby incurring a deficit and requiring a public subsidy). Coase also criticized the absence of a market test to determine whether users are willing to pay the full cost of supplying the commodity; the redistribution of income to favour users of decreasing-cost industrial products; and the impetus towards centralization of the economy.

Most of these criticisms can be dealt with by a multipart pricing system: the first part sets marginal price equal to marginal cost while the second part levies an assessment to recover those costs that exceed marginal costs. Even so, multipart schedules often fail to reflect the economic concept of opportunity costs correctly, focusing instead on recovering historical or embedded costs. The relevant opportunity costs include both the cost of securing incremental supplies of water and the value of water in alternative uses. Opportunity costs should be determined after adjusting prices to allow for distortions brought about by government interventions in pursuit of other objectives. In economic jargon, "shadow prices" must be used.

The average cost pricing principle calls for recovering all costs by charging for each unit according to the average cost of providing all units. It is simple and easy to understand, as well as fair and equitable. Beneficiaries pay only the resource costs incurred on their behalf. The desired signals to users are provided, although not in as precise a way as with multipart pricing. Here, too, often only historical costs, not opportunity costs, serve as the basis for calculating average costs.

The ability-to-pay principle rests heavily on the equity criterion. Water charges are dependent on income or wealth rather than on costs. This principle is the most common basis for setting irrigation rates throughout the world and is also regularly applied to village water supplies in developing countries. Economists who view water as a commodity tend to be critical of the ability-to-pay approach. Since charges bear little relation to costs, no allocative test of willingness to pay is provided. This ability-to-pay concept is inherently subjective and political pressures frequently influence the formula in ways that distribute wealth from taxpayers to water users.

In many places throughout the world, water is scarce enough to justify the tangible and intangible costs of establishing formal pricing systems. Flat rates could satisfy cost repayment requirements in the absence of serious shortages. However, when the signals of water scarcity are absent, pressures arise for structural solutions (more construction to capture, store and deliver water) to satisfy incorrectly perceived water "needs".

The inevitability of scarce water supplies suggests the eventual adoption of multipart rate schemes that reflect the real or opportunity costs of water and other resources required for service provision. The literature describing the most desirable form for water markets and the literature dealing with water pricing have converged on the notion of a pricing system that reflects the opportunity cost of water via the mechanism of transferable water entitlements.

Coordinating Groundwater Extraction

Groundwater is an extremely important resource for many developing countries, including Bangladesh, India, Pakistan and the entire Near East region. In India, tube wells accounted for nearly one-half of the net irrigated area by the late 1980s.48 Because it deals with the complex interaction between society and the physical environment, aquifer management presents a formidable problem of policy design. Two types of collective policy decisions must be addressed in the management or regulation of overexploited aquifers. For one type, termed "managing the water", decisions are based on: (i) the appropriate annual rate of pumping; (ii) the geographic distribution of pumping; and iii) whether water supplies are augmented and/or the aquifer artificially recharged. The other type of policy decision, "coordinating the pumpers", determines: (i) the institutions and policies that divide the extraction rate among potential individual users and user classes and that influence pumper behaviour; and (ii) how rules for limiting pumping are monitored and enforced.

The three broad types of institutional arrangement for managing aquifers are prices and charges, quantity-based controls and exchangeable permits.

Prices and charges to control pumping. Charging pumpers is one potential method of achieving economically efficient extraction rates. An appropriately scaled charge or tax confronts pumpers with both the foregone user cost and the external cost (from increased pumping costs) imposed on neighbouring pumpers. This type of water charge internalizes user costs and external costs and achieves an optimal extraction rate.

In aquifer management, this approach takes care of one important difficulty-pumpers impose costs on themselves (that is, the external costs are reciprocal). The reduction in water use resulting from a tax would be at the expense of redistributing rents to the taxing authority, thus lowering the net income of the aquifer exploiters.

Quantity-based controls. Quantity-based control mechanisms range from simple well permits to exchangeable pumping entitlements. Well and pump permits grant the right to install and operate a well of a particular capacity. Irrigation permits frequently specify the lands on which well water can be used, thereby restricting the transport of water to other sites. To protect existing pumpers, permits for new wells may restrict locations. For example, the state of Colorado in the United States identifies "designated groundwater basins" (for aquifers with limited natural recharge) in which new well permits must meet specific criteria-no more than 40 percent of stocks may be exhausted within a three-mile radius over a 25-year period. In most cases, well permits do not set limits on the quantity of water pumped. The economic limitations imposed by pumping costs and crop prices are assumed to be sufficient to inhibit excessive withdrawals.

Permits with appropriate size and spacing specifications can slow extraction rates. They are relatively easy to monitor and are reasonably palatable to pumpers who strongly reject more stringent regulatory devices. On the other hand, permits are most effective before problems have become severe and complex-in cases when preventing new wells and pumps solves the problem or when pumped water is not exported away from the area overlying the aquifer. In more serious cases, where all existing users must reduce annual extractions, regulating rates of withdrawals must be considered. A pumping "quota" is a more precise quantity control mechanism. The quota specifies a fixed annual rate of extraction for each water user. The initial quantity might be assigned in proportion to use in a base period (although such an approach might set off a pumping race to establish initial rights) or be based on the proportion of land that is owned overlying the aquifer. The technology for metering withdrawals is neither complex nor expensive so, if the pumpers are willing to be metered, regulatory monitoring and enforcement need not be difficult. In principle, pumping quotas are no different from conventional surface water rights, which entitle owners to fixed shares of each year's available flow.

Anecdotal evidence suggests that farmers who have previously enjoyed an unregulated aquifer believe they are entitled to unlimited withdrawals for use on lands overlying the aquifer. They are frequently reluctant to submit to metering and the meters, once installed, are reportedly subject to high rates of unexplained "breakdowns".

Very small wells for livestock or individual households could be exempted from the permit and quota system. At moderate levels of overdraft, the cost of monitoring every small pump set might outweigh gains from reduced pumping. Also, for income distribution reasons, policy-makers may not want to restrain smallholders.

Transferable pumping entitlements. When a fixed quota is too inflexible in the face of changing water stocks and demand conditions, transferable pumping entitlements are an alternative. The pumping entitlement can be divided into two parts: one component may provide a claim to the stock of water and the other to the annual recharge. Both claims may vary from year to year, with allotments set by the groundwater authority. Annual rights to the basic stock would vary according to current and anticipated economic and hydrological conditions (including energy and commodity prices, interest rates and the remaining stock of groundwater). Rights to the natural recharge and return flows from human uses could be set to reflect a moving average of estimated recharge in recent years.

The transferability of entitlements promotes economic efficiency over the long term, permitting a reallocation to higher-valued uses as economic conditions change. Transferable rights are also consistent with local control criteria and require minimal interference with individual freedom to operate a farm or business enterprise. One recent study suggests that tradeable rights and water markets for groundwater appear to be increasing in India, where as much as one-half of the gross area irrigated by tube wells involves purchased water.

In groundwater management, the quantity-based approaches appear to be preferable to pumping charges. They can yield economically efficient solutions with simpler monitoring and enforcement burdens, while avoiding the redistributive implications of taxes or subsidies. While some new outside controls on pumping are required, they need not be any more repressive than property rights for other resources or commodities.

Conjunctive Groundwater and Surface Water Management

Joint management of interrelated stream-aquifer systems is called conjunctive ground-surface water management. Aquifers interrelated with flowing streams frequently present both distinct management opportunities and problems. Unrestricted access to groundwater may reduce the water available to those holding rights to streamflows. In the state of Colorado in the United States, a problem arose two decades ago for irrigators in the South Platte basin. Groundwater exploitation reduced streamflow, but by only a small fraction of the amount pumped. The most obvious option-placing the pumpers into the existing surface water rights system while protecting those holding existing rights to surface water-would have sacrificed most of the substantial economic benefits of exploiting the aquifer.

After several methods had been experimented, a solution was found based on markets for existing rights to surface water. In the event of a shortage, groundwater users could replace the portion of streamflows taken by pumping the aquifer. They could also replace water by purchasing and delivering rights to reservoir water. Young, Daubert and Morel-Seytoux demonstrated that this decentralized approach is economically superior to the alternative of forcing pumpers into the surface water rights system. The ready availability of substitute water supplies and the existence of flexible water transfer institutions are necessary to implement a solution of this type. Numerous opportunities for this market-based approach do exist, however, in the large alluvial basins of the Indus and Ganges-Brahmaputra.

Preserving Water Quality

Human production and consumption activities generate pollution by extracting and processing raw materials into consumer goods. Some wastes (residuals) from the production process are returned to the environment (e.g. waste chemicals from petroleum refineries discharged to rivers). Similarly, households return unwanted by-products of consumption activities to the environment-to sewers, to the air or to sites receiving solid waste. The materials balance principle, derived from basic laws of physics regarding the conservation of matter, asserts that, over the long term, the mass of residuals discharged to the environment must equal the mass of materials originally extracted from the environment to make consumption goods. The environment's importance as an assimilator of residuals is equal to its importance as a source of materials.

The Water Sellers

One does not have to be a landowner or even a farmer to benefit from irrigation. Irrigation increases employment and provides the landless with opportunities to work on farms or in upstream or downstream activities. In Bangladesh irrigation has also opened profitable new avenues for the landless by enabling them to exploit and sell water. The "water sellers" are organized with the help of PROSHIKA, one of a number of NGOs that aim to develop an irrigation service for farmers by tapping the abundant groundwater which underlies much of Bangladesh.

Fresh groundwater is a widespread resource but is usually present in small amounts, only sufficient to service household needs. However, in the great alluvial basins such as the Nile, the Indus and the Ganges-Brahmaputra, the alluvium may be 100 m or more deep and 10 or even 20 percent of its volume may be freshwater. The vast reserve of groundwater is recharged annually by floods, canal and field seepage and rainfall infiltration.

Groundwater is particularly valuable because it is available consistently and, unlike surface reservoirs, evaporation losses are minimal. Where surface canal supplies are also available, and provide a more or less constant base supply, groundwater can be used conjunctively to satisfy the peak demands of crops. In addition, groundwater is usually available close to farms and is more under the control of farmers.

In rural Bangladesh more than 50 percent of the population is landless or has less than 1.2 ha of land. Providing the poor access to productive resources such as water is clearly important. The PROSHIKA experience was based on organizing landless groups, using credit effectively to purchase mobile pumping equipment and providing a reliable service to farmers and share-tenants.

The water sellers targeted their service in areas where farmers had very small scattered plots of land and irrigated their fields from shallow low-cost boreholes, using portable diesel pump sets. The farmers pay the sellers in crop share, cash or, occasionally, a fixed amount in kind.

The success of the PROSHIKA mission depended primarily on access to credit (which PROSHIKA helped to organize) as well as skills training and technical support relating to agriculture, management, literacy, health and group solidarity.

The water sellers: improved water-use efficiency and equity by improving the direct access of small farmers who are usually tail-enders in other systems of irrigation; benefited from the more equitable distribution of productive assets between those with and those without landholdings; created additional employment within and outside the group as a consequence of more productive agriculture; obtained cash to buy more food and promote commercial agriculture; participated in the developing water market, which has prevented richer peasants or landlords becoming monopoly "waterlords"; and showed that the poor can be creditworthy without land as collateral.

The PROSHIKA experience is being replicated throughout Bangladesh and has many lessons that may serve for other developing countries.

Source: G.E. Wood, R. Palmer-Jones, Q.F. Ahmed, M.A.S. Mandal and S.C. Dutta. 1990. The water sellers: a cooperative venture by the rural poor. West Hartford, Connecticut, Kumarian Press.

One important policy implication of the materials balance principle is that residuals must end up somewhere, either as mass or energy. The management of discharges into watercourses must be integrated with waste disposal into the atmosphere and landfills. Reducing the amount of waste discharged into water may not solve society's overall problem if the waste is sent elsewhere, such as into the atmosphere by burning or to a land site by dumping.

There are two types of water pollution: point source and non-point source. Point source pollution refers to cases where a readily identifiable source, such as a pipe or ditch, transports the pollutant to a water body. Regulation and monitoring focus on the point of discharge. In non-point source pollution, no single source of pollutant discharge is easily identifiable but the collective effect of numerous sources results in a significant impact. Non-point pollution problems pose a difficult and costly management challenge.

NON-POINT POLLUTION CONTROL OPTIONS

Policy options to control non-point water pollution present special difficulties because of the great variety of sources and pollutants. The primary source of non-point pollutants is the agricultural sector. Fertilizers and pesticides are carried off the soil surface into lakes and streams or percolate into groundwater deposits. Aquifers become polluted by nitrates from fertilizer application and livestock wastes. Timber harvesting, land clearing for urban development and mining also originate non-point pollution of waters. Urban storm drainage, leakage from buried fuel tanks and subsurface and surface mining are other contributors. Runoff from farms and forests may carry suspended solids and sediments, dissolved solids and chemicals (mineral fertilizers, particularly nitrogen and phosphorous, and pesticides). Other substances that often occur in diffuse source runoff are oxygen-demanding organic matter, petroleum products, heavy metals and faecal bacteria. Non-point source pollution is also characterized by its episodic nature. Occasional heavy rainfall or snowmelt is typically the trigger, in contrast to the more even flows of discharge from point sources. These characteristics of source type and timing imply that a variety of control technologies may be required for effective abatement.

Non-point pollution control may also be determined by the nature of human activities causing the problem. For example, the pollution resulting from a farmer's land depends not only on the rainfall patterns and the land characteristics (slope and soil texture), but on numerous prior land-use and production decisions, including choice of crops, tillage practices and pesticide and fertilizer use. The farmers' production choices are, in turn, influenced by market prices for inputs and products as well as by government price and income support programmes. In fact, pollution from the farm sector is exacerbated by government policies that make certain crops overly attractive. Successful policy interventions must change those aspects of farmers' decisions that are the source of pollutants.

Policy options for non-point pollution control are classed as cognitive, regulatory and incentive-based. Cognitive (voluntary) approaches use education, moral persuasion and technical assistance to influence the behaviour of polluters. Cognitive approaches are attractive because of their low economic and political costs. They have been tried in some countries but have had limited success. Several factors account for this; for example, private costs incurred to change land-use practices can be substantial while private gains may not be obvious.. Because of the uncertain linkage between changing production decisions and improving water quality (often at distant locations), individuals have little incentive to try new approaches.

Regulatory policies call for specific actions or prohibitions against those responsible for water quality degradation. One approach is to use "design standards" that specify actions to be taken (such as a management plan for sediment control) or actions prohibited (such as certain cropping practices on highly erodible lands). "Performance standards", in contrast, place limits on the rate of pollution discharge to a water body. In this case, interference with land-use practices is only in response to observed violations. Neither technique is without limitations. Design standard regulations are easier to enforce; however, they may be unnecessarily costly because their general application may impose costs on those who contribute little to the problem. Performance standards, in principle at least, focus more directly on the pollutant source but are difficult to monitor and enforce. Because an accurate measurement of discharges (particularly from small farms) is nearly impossible, disputes over actual sources of pollutants are unending.

The alternatives to regulatory policies include various incentive methods such as taxes, subsidies and emission trading policies.54 Taxes or fees can be levied on either inputs or pollution outputs. For example, extra charges have been imposed on agricultural fertilizers in Sweden, with proceeds used to fund water quality monitoring. Higher costs are expected to reduce fertilizer application rates and, therefore, water pollution. However, taxes are unlikely to be set high enough to affect land use significantly because of the adverse effects on income.

Alternatively, charges may be levied for pollution by imposing an "effluent charge". However, the technical and administrative complexity of setting fees and linking numerous farmers precisely to the damages caused by their effluent is mind-boggling. No successful example of this type of taxation of non-point source pollution is presented in the literature.

Subsidies could encourage farmers to reduce pollution, adopt more appropriate land-use practices or make environment-friendly investments. Subsidies to prevent soil erosion (and the associated productivity losses) have a long history in many countries and are the most politically attractive of the available options. In contrast to other approaches, which impose costs on the emitting source and spread benefits over the entire society, subsidy costs are spread over the general population and gains are offered to the land user. Nonetheless, paying polluters to avoid polluting activities remains objectionable to some groups. Moreover, payments may be made to individuals who would adopt proper practices anyway.

Finally, the outright purchase of water rights and/or land-use rights is another approach. For instance, a public agency could acquire rights to part or all of the polluting lands and manage them to safeguard water quality. The purchase of tropical forest lands by either public or private agencies has been undertaken to preserve first-growth forests, with water quality improvements as a side-benefit. Again, costs are borne primarily by beneficiaries rather than by the land users whose practices are actually responsible for the pollution.

POLICY ISSUES IN IRRIGATED AGRICULTURE

Irrigation is an essential component of sustainable agricultural development but it is not a unique sector, since it faces challenges similar to those confronting other public and private sector economic activities. The previous two sections (II and III) of this chapter examine how various demand-side policy measures can help shape decisions that encourage water-use efficiency. However, while appropriate policies and regulations are necessary for improved water productivity, a variety of additional water-saving measures are required in the irrigation sector.

Some water-saving measures involve taking more advantage of the scientific, engineering and technological advances in soils, plants and irrigation. Other measures focus on administrative and managerial reforms to improve efficiency, including the decentralization of public irrigation agencies and a greater reliance on farmer-owned and farmer-operated irrigation.

This final section highlights three key irrigation issues: declining growth and investment trends in irrigation; the difficulties imposed by irrigation-induced environmental degradation; and efforts to reform managerial and administrative systems. Many of today's irrigation-related problems appear imposing and even overwhelming. The purpose of this section is not to present a discouraging overview of irrigation's future but to focus on important issues that will shape its future.

When scarce water is under human control in irrigation systems, there are many opportunities to use it optimally. Understanding the problems associated with improper irrigation as well as the potential for efficient irrigation is a first step in the search for these opportunities.

Irrigation in the 1990s and Beyond

Many pressing irrigation issues reflect the various economic, social and political influences affecting society in this period of changing economic conditions and environmental concerns. At times, irrigation-related problems are the result of a distorted macroeconomy which, despite providing operating subsidies, renders farming unprofitable and results in repeated underinvestments on farms over long periods. If the macroeconomy is performing poorly and if prices in the economy are markedly distorted from their real value to society, then it is inevitable that irrigation will suffer.

In other cases, irrigation itself is the problem. The overall performance of many irrigation projects is disappointing. Evaluations document a wide range of problems, including: cost and time overruns; poor management; the non-realization of full, planned benefits; adverse environmental and health impacts; and the exacerbation of inequities in the existing social and economic distribution of assets among farmers.

Trends in Irrigated Area

In 1800, approximately 8 million ha of farmland were irrigated around the world. By the end of the nineteenth century, the irrigated area had expanded to 48 million ha, mostly as a result of large water projects in India and what is now Pakistan. By 1990, net irrigated area reached 237 million ha, of which nearly three-fourths were in the developing countries. China, India and Pakistan alone now account for about 45 percent of the world's irrigated area and 60 percent of the developing country total.

On a global basis, irrigated area expanded by an average of 1 percent per year during the early 1960s, reaching a maximum annual rate of 2.3 percent from 1972 to 1975. The rate of expansion began decreasing after 1975 and is now less than 1 percent per year. At current levels of population growth, the slower expansion in irrigated areas is resulting in an unprecedented decline in the per caput amount of irrigated land. Among the reasons for the decrease are increased construction costs, falling real prices for wheat and rice, a growing awareness of environmental and social costs and poor irrigation performance at the farm and project levels.

Crop Prices and Construction Costs

Construction and operation costs for irrigation projects have risen steadily over the past four decades as the world's best land and most of the readily available water supplies have been developed. Over the same time period, world cereal prices fell sharply.

For example, the real price of rice fell by 40 percent between the mid-1960s and the late 1980s.

In Indonesia and India, the real costs of irrigation have doubled since the early 1970s. In the Philippines and Thailand, irrigation construction costs have increased by 50 percent and in Sri Lanka, costs have tripled. After reviewing various recent reports and evaluations, including many from the World Bank and FAO, Postel reports that: "Today, capital costs for new irrigation capacity run between $1 500 and $4 000 per ha for large projects in China, India, Indonesia, Pakistan, the Philippines and Thailand. They climb toward $6 000 per ha in Mexico.

In Africa, where roads and other infrastructure are often lacking and parcels that can be irrigated are relatively small, per hectare costs have climbed to $10 000-$20 000, and sometimes even higher. Not even double-cropping of higher-valued crops can make irrigation systems at the top end of this spectrum economical." It is not just large-scale projects that have become so expensive; FAO estimates that even medium-sized irrigation construction costs range from $2 400 per ha in

Asia to $7 200 in Africa. Likewise, the modernization of existing irrigation projects is becoming increasingly expensive. Many old projects designed for monocropping have produced gradually declining yields. These systems need to be redesigned to permit crop diversity, increase yields, conserve water and reduce environmental hazards. Modernization involves canal lining, improved hydraulic control structures, better land development and appropriate irrigation techniques.

Efficient Water use at the Farm Level

The current and potential roles of rain-fed and irrigated agriculture need to be quantified, taking into account the probabilities of rainfall, the available irrigation water supplies and their cost. Small-scale irrigation, including supplementary water for rain-fed agriculture, and a variety of water harvesting and water spreading techniques have considerable potential and should be developed further.

The proposed priority action plan points out that increases in developing country production in the 1990s must come first from increases on existing irrigated lands and second from increases on rain-fed lands. In order to improve performance and control groundwater levels in irrigated agriculture, it is necessary to establish monitoring, evaluation and feedback systems. Appropriate irrigation management training must be available and irrigation extension services must be strengthened and developed further. Information must be shared among farmers, extension workers, design engineers and researchers for a better understanding of various approaches and techniques. Measures should also be taken to review, develop and implement water pricing policies, establish effective demand and supply management procedures as well as cost recovery mechanisms for the operation and maintenance of irrigation projects.

In order to increase production on rain-fed lands, it will be necessary to employ existing soil and water management knowledge to help increase the efficient use of agricultural water. Additional research in the areas of rain-fed water and soil management is also necessary and examples of successful and effective rain-fed practices must be collected and disseminated.

Waterlogging, Salinity and Drainage

Waterlogging and salinity are among the principal causes of decreasing production on many irrigated projects. Waterlogging is due to an excessive input of water into systems that have finite natural drainage capacities.

After waterlogging has occurred, soil salinity increases because the irrigation water leaves dissolved solids in the soil. It is essential to monitor the water table levels from the beginning of a project in order to implement corrective measures before soil damage has occurred.

The action plan notes that, in rain-fed agriculture, surface drainage is required to prevent any temporary waterlogging and flooding of lowlands.

In irrigated agriculture, artificial drainage is essential under most conditions. It is vital to minimize drainage requirements and costs by reducing the sources of excess water through improved system design and on-farm water practices. The implementation of groundwater monitoring and water balance studies will help predict drainage requirements. Soil salinity monitoring in problem areas should also be started in order to adopt practices to overcome the problems. In addition, pilot drainage projects should be established in waterlogged and salinized areas to verify the design and effectiveness of materials.

Water Quality Management

In sustainable agricultural development, there are two significant aspects of water quality: the quality of water used in irrigated agriculture should not cause crop damage; and agricultural activities should not adversely affect the quality of surface or groundwater in such a way that their subsequent use is limited.

The action plan suggests that groundwater quality monitoring programmes to assess agricultural water quality and its impact be developed and that strategies be implemented to minimize the extent of water pollution inflicted by agricultural activities. In addition, national strategies and plans should be prepared, outlining the rational use of treated waste water and drainage water for agriculture.

Small-scale Water Programmes

Small-scale projects are normally undertaken by communities or individuals who develop and operate most activities of the projects themselves, although some technical assistance is often necessary. Small-scale programmes include a diversity of technologies such as water harvesting, well development, river offtakes and use of wetlands.

The action plan submits that small-scale water programmes can assist sustainable agricultural development. However, further expansion must be founded on adequate technical advice and support, improved institutional collaboration and greater involvement of the local communities. National policies and programmes should be developed for implementing small-scale water projects for rural development. Accordingly, it will be necessary to enhance the capability of farmers to implement, operate and maintain small-scale water programmes.

The monitoring and evaluation of the small-scale water project will help identify the failures as well as the successes.

Scarce Water Resources Management

Many countries face constraints to their development, as available water resources are insufficient to meet demand. The action plan notes that water scarcity conditions require long-term strategies and practical implementation programmes in order to ensure that agricultural water is used in ways consistent with limited water resources.

A strategy must be formulated for the management of land and water for sustainable agricultural development under water-scarce conditions.

At the same time, these policies must remain compatible with the prevailing socio-economic conditions as well as the environment. In addition, measures should be implemented to deal with drought situations and to improve community resilience during such times.

Supporting Action

The five areas outlined above require common supporting actions, i.e. the development of adequate databases, adaptive research, institutional strengthening, human resource development, improvements in socio-economic analysis, environmental protection, technology transfer and infrastructure development.

Increasing real investment costs (including recognition of environmental costs) and falling crop prices have resulted in a substantial reduction in new irrigation projects. Aggregate lending for irrigation by both international donors and the major cereal-producing countries in the 1980s was only half what it was in the 1970s. This reduced investment has, in turn, resulted in falling growth rates in irrigated area.

Irrigation and Land Degradation

Escalating demand pressure on water resources during the past decade is only one of the concerns for policy-makers. Both watersheds supporting water resources and the land base supporting irrigated agriculture are becoming degraded. Industrial and domestic pollution are affecting irrigated agriculture, while sedimentation washed from denuded hillsides is filling up irrigation reservoirs and small tanks. At the same time, improper irrigation practices are causing waterlogging, salinization, soil erosion and water pollution which directly affects other irrigators.

FAO estimates that of the 237 million ha currently irrigated, about 30 million ha are severely affected by salinity and an additional 60 million to 80 million ha are affected to some extent. UNEP recently reported that the rate of loss of irrigated land from waterlogging and salinity is 1.5 million ha per year.62 Millions of hectares of irrigated land, from Morocco to Bangladesh and from northwestern China to central Asia, suffer from this progressive condition. Salinity-affected areas as a percentage of total irrigated area is estimated to be 10 percent in Mexico, 11 percent in India, 21 percent in Pakistan, 23 percent in China and 28 percent in the United States.

Salinity is caused by a combination of poor drainage and high evaporation rates which concentrate salts on irrigated land; it mainly occurs in arid and semi-arid regions. Even good-quality irrigation water contains some dissolved salt and can leave behind tonnes of salt per hectare each year. Unless this salt is washed down below the root level, soil salinity will result. A number of factors influence salinity, including the water table depth, the capillary characteristics of the soil and management practices regarding the amount of water applied in excess of plant evapotranspiration to leach the salts.

A related concern is the rapid rise in groundwater levels, leading to waterlogging and depressed crop yields. Waterlogging is not an inevitable result of irrigation. It occurs when excessive water is used in systems with finite natural drainage capacities. Seepage occurs if: soils are very light; canals and watercourses are not lined or maintained; farmers near the head of a system withdraw or apply excessive amounts of water; fields are not levelled; and/or the delivery system cannot respond to rainfall by closing inflows.

Irrigation can raise groundwater levels to within a metre or so of the surface, which results in secondary salinization when the groundwater brings to the surface dissolved salts from the aquifer, subsoil and root zone. If seepage and horizontal recharge exceed evaporation and natural drainage, then groundwater levels rise, eventually causing waterlogging. In arid areas where the upward movement of water and evaporation exceeds downward percolation and where the groundwater, soil or irrigation water contains some salt, the buildup of salt in the soil surface layers will eventually reach toxic levels.

Countries suffering from waterlogging and salinity face a dilemma. They cannot force abandonment of affected lands because of the growing populations that depend on them, but neither can they afford to drain the lands. The cure for rising water tables is drainage and improved water management to reduce percolation, but drainage is expensive and improved water management requires both on-farm investment and the training of extension personnel and farmers.

At the end of the 1950s, Pakistan began to sink wells to pump out salt-laden water. The initial capital costs were high and operating costs have been increasing constantly. Between 1971 and 1985 the cost of managing the wells and drains rose fivefold. Today, Pakistan spends more on reclaiming land than on irrigation. The cost of maintaining drains is five times the recurrent costs of supplying water and most farmers pay only half the costs of delivering irrigation. Some policy-makers argue that, as the value of water rises, management will improve and much of the wastage (which creates unhealthy conditions as well as increased drainage requirements) will be reduced. Farmers may then be willing to pay for a drainage service that makes their investments sustainable.

Irrigation: Good Government and Good Management

Political pressures often prevent the implementation of apparently sensible and fair reforms for water services. Where water is regarded as a special commodity or has emotional or religious importance, governments are reluctant to charge farmers for irrigation. Policy-makers often find it extremely difficult to raise sufficient revenue to match even their priority needs. The practical effects on traditional public service activities may be harsh. Water has been one of the first sectors to feel the effects of budget-saving efforts and changing resource availability. However, it is unlikely to be treated more austerely than other areas. Box 19 discusses the relationship between good government and irrigation performance.

A prime opportunity for irrigation advancement-and indeed for development in general-lies in the enormous potential of the 237 million ha already operating. While the total value of irrigation investment in the developing world today is about $1 000 billion, the returns are well below the known potential. Many irrigation schemes need substantial investment to be completed, modernized or expanded. Although it is increasingly expensive, rehabilitation can yield high returns.

Good Government and Irrigation Performance

Improvement in irrigation performance depends on good government, or governance. This may be an obvious assertion, but what exactly does the term mean for irrigation? There are four main elements of governance which can be considered at the national or the local level: the legitimacy of government; its accountability, its competence; and its respect for human rights and the rule of law. Legitimacy refers to the way in which a population gives consent to be governed, how they are consulted and whether the consent can be withdrawn. Accountability of politicians and officials is tested by how they explain their role and decisions, provide information and can be held responsible for their behaviour. A government demonstrates competence in formulating policies and translating them into action in a timely and effective way. Governments who respect human rights establish a framework of known laws, applicable to all, without bias or corruption, with limits on and protection against the exercise of arbitrary power. We can illustrate how these four elements of governance might affect irrigation with simple examples.

Legitimacy. When a new project is planned, are those living in the area consulted about the design of the scheme? Are there recognized representative groups of farmers, including women? Are the office holders elected and accountable to the members? Do these groups participate in decisions that affect them? For example, if a groundwater irrigation project threatens the availability of drinking-water from hand pumps, are self-help groups informed and invited to make representations?

Accountability. Are the financial plans of the irrigation scheme made public and arrangements made to explain them to farmers? Are there performance criteria with audit arrangements to ensure that officials adhere to the rules and, if they fail to perform satisfactorily, call them to account? Are officials responsive?

Competence. Can the professional staff prepare accurate budgets and effectively deliver services such as timely canal maintenance? Are there arrangements for training them or replacing them with competent officers if they fall short of their duties?

Rule of law. Is there a clear legal framework to regulate groundwater abstraction to prevent overpumping of the aquifers? Is it enforced? Can pollution by industry or by saline water from upstream drainage projects be regulated? Are illegal extractions by farmers at the head of canals monitored and offenders charged by legal processes that are fair, timely, objective and without discrimination on grounds of race, gender or minority status?

Beyond capital investment, investment in human resources is needed to improve economy, efficiency and effectiveness in management.

Investment in irrigation management can be an integral component of a productive rural employment policy and resource management can be improved to match developments in engineering and science. Those concerned with irrigation development and the welfare of farmers and their families are beginning to take a broader view of their responsibilities, paying as much attention to management and development as to engineering.

A review of recent literature on public sector policy and management suggests the need for transforming the typical irrigation authority from a technical, engineering and construction culture to an organization that is focused more on people (farmers) and that is more autonomous and more responsive to customers, i.e. a strategic management service agency.

In the developing world, many water subsector groups recognize their role in the management of valuable water and other complementary resources for economic growth and poverty alleviation.

Irrigation management does not generally have a good record, but the full extent of the problems has been obscured by a lack of performance assessment criteria and a variety of hidden subsidies. It has often been assumed that, if farmers are not complaining, then management must be good. A lack of complaints in a highly subsidized system is not a good indicator of efficiency. Similarly, managers with obvious problems have sometimes had to cope as best they could with a range of exogenous factors that limit their effectiveness, for instance overvalued exchange rates.

One example of a remarkably successful administrative reform is found in the Philippines, where irrigation operating expenses have been reduced, revenue has matched operating costs and operational performance has improved in the process.

Water user Associations and NGOs

An increasing number of private sector groups, including water user associations and other NGOs, are taking over some public sector irrigation responsibilities. The inclusion of water users in irrigation planning, management and ownership is proving to be an effective method for increasing irrigation system efficiency in many cases. Studies throughout the world demonstrate that user participation in irrigation services improves access to information, reduces monitoring costs, establishes a sense of ownership among farmers and increases transparency as well as accountability in decision-making.

Water user associations are expected to increase in number and importance over the next decade as the stress on self-reliance increases. Already, governments are turning many aspects of public irrigation systems over to water user associations. Well-documented examples can be seen in Argentina, Colombia, Indonesia, Mexico, Nepal, the Philippines, Sri Lanka and Tunisia. In Indonesia, for example, the government had transferred more than 400 irrigation systems, covering 34000 ha, to water user associations by 1992. In the future, as farmer financing becomes more commonplace, user groups will become even more powerful.

Other NGOs undertake a wide range of water-related functions, from developing projects for rural water supplies and minor irrigation to fostering water user associations for water management purposes. Some NGOs encourage farmers to try new technologies, for example the catchment protection and sprinkler irrigation techniques introduced by the Aga Khan Rural Support Programme in Gujarat, India.

Many NGOs stem from local initiatives and operate as independently funded and self-managed groups. These organizations bring fresh views, new ideas and participatory working methods to other areas of development policy and practice. Much of their success is attributed to their local knowledge as well as their interest in and experience of regional conditions. They have been particularly active in promoting the interests of poor and disadvantaged groups through articulate and forceful advocacy and service provision. In addition, the local base of NGOs may allow them to reach vulnerable or remote groups which are exceptionally difficult to reach with conventional public schemes.

Water is not an easy sector in which to promote cooperation, but the potential gains are high, which makes renewed efforts worthwhile. Resolution of many water allocation and development problems requires a common willingness to forego personal benefit for the social good. Government efforts to promote personal sacrifice through economic policies, laws and regulations that require self-restraint, such as water rationing or optimum groundwater pumping regimes, have seldom proved effective. On the other hand, with their close local contacts and skills in group mobilization and cohesion, NGOs can provide the institutional leadership required to bring about socially optimum solutions.

Self-financed Operations: The Philippines' Bureaucratic Reform

The National Irrigation Administration (NIA) of the Philippines is a good example of how a bureaucracy can, over time, transform its strategy and operating style. Since the mid-1970s, the NIA has evolved from an agency that focused primarily on the design and construction of irrigation systems and told farmers about its key decisions into an agency that gives priority to the management and maintenance of irrigation systems, and that gives farmers, through their membership in irrigators' associations, the opportunity to participate in irrigation system management and to make key decisions about maintenance. How did this transformation come about and what have been its effects? Commitment from the top. In the early 1970s, the NIA's top management diagnosed correctly that farmer organizations were crucial to effective irrigation management and, therefore, committed the NIA to developing and extending the responsibilities of irrigators' associations. The NIA's willingness to renounce its traditional authority was a key factor in this process.

Legal and financial status of the NIA. In 1974, the NIA was established as a public corporation and ceased to be part of a government ministry. It was given a five-year lead-in period in which to become financially self-sufficient in terms of its operating budget. The NIA's semi-independent status set the scene for farmer-financed irrigation and the devolvement of management tasks to farmers.

A gradual approach. The NIA did not transform itself overnight. The participative approach was first tried in the mid-1970s with smaller-scale "communal" irrigation systems that were traditionally managed by farmers. The end of the 1970s saw the development of a methodology to maximize farmer participation based on two pilot projects. Lessons learned were then incorporated into plans to manage the large-scale "national" irrigation systems jointly. Almost 20 years on, the transformation of the NIA from a top-down bureaucracy continues.

Understanding farmers. How has the NIA motivated farmers to participate in and pay for irrigation, to commit time to maintenance, to liaise with the NIA and to plan for the future? The NIA experience illustrates some important preconditions for farmer participation: teaming community organizers and engineers in order to integrate social and technical activities into one process; involving farmers in all project activities from the very beginning, thereby building up their organizational skills; modifying NIA policies and procedures that obstruct farmer participation; allowing enough time for farmers to mobilize and organize themselves before new construction activity.

The quality of the irrigation service provided by and for farmers has undoubtedly improved, system operating expenses have been reduced and the recurrent cost burden of the scheme on the national budget has been eliminated. But what about the impact on irrigation performance? Recent research suggests that the reforms have made water supply more equitable.

On five schemes surveyed five years following the cessation of subsidies, the main impact was the increased equitability of supply. For reasons unconnected with the management reforms, water supplies fell by 13 percent but crop area held constant and yields remained the same. This confirmed the connection between equity of supply and financial viability. To appreciate the value of the reform we need to consider what would have happened to the deteriorating service if the impoverished state bureaucracy had continued to run the schemes.

FUTURE DIRECTIONS IN WATER MANAGEMENT POLICY

Sustainable agricultural development depends on sustainable water use. Governments today recognize that the search for sustainable economic growth requires, in part, both economy-wide and sector-specific policy reforms. Economy-wide policies attempt to create a favourable macroeconomic environment while water sector policies, for example, seek to encourage resource efficiency among water users.

The current emphasis on macro-economic policy reforms and economic liberalization has several important implications for irrigation. One of the most important is that the era of large direct and indirect subsidies is nearly over. Moreover, recognition of the value of water (and the high cost of turning a water source into a service delivered to a farm) makes the water sector a prime target for further policy reforms. Nonetheless, irrigation remains a resource-hungry sector in this transitional period. Even successful irrigation consumes large quantities of capital and foreign exchange and ties up scarce skilled personnel.

Like many public sector personnel, irrigation managers must walk a fine line between a tighter control of finance, the need for more positive active leadership and better planning of resource allocations, on the one side, and the contradictory need for more ideas from below (farmer customers), on the other. Financial pressures are likely to be the dominant influence. Irrigation as a public sector agency still relies on budget allocations to obtain financing. Many argue that this gives little incentive to save money and may, in fact, have the reverse effect.

As private sector disciplines are applied in irrigation, policy-makers are finding that: agencies become more supportive of farmers' own efforts and less inclined to make all key decisions before informing farmers accordingly; management seeks more consensus on priorities, more information about the basis of decisions and a common view of external factors affecting management; irrigation schemes seek and receive more autonomy; the financial responsibilities and accountability of managers increases; and managers shift focus from their ministries and governments, depending on the amount of finance generated by service fees. At present, water engineering feeds off mathematics, physics, chemistry and biology; management feeds off economics, psychology, political science, history and philosophy and also has a high requirement for personal skills in communication, negotiation and teamwork. However, as changes occur in irrigation in many countries, managers find they must acquire new knowledge and a new set of skills.

Following are some of the issues emerging:

- National water politics are shifting from projects to policies-this trend is likely to continue and even accelerate.
- Water may become a test-bed for economic reform, liberalization and accountability.
- Given water's scarcity and its value to cities and industry, the water subsector will be less dominated by irrigation and its multipurpose uses will be more widely acknowledged.
- Irrigation is a service with customers and users; it is not a production industry.
- At the level of the irrigation scheme, the process of water policy formulation, assessment and appraisal needs to include more open groups that are representative of political, technical, managerial and (most important) water user associations.
- These policy groups would be consulted before policy selection and then provide feedback and adjustment in the light of experience.
- The policy groups would identify options consistent with the national policy framework, as opposed to measures to protect and satisfy special irrigation interests.

- The goal is to identify a broader range of water policy options, to have less "policy by crisis" management and more resilience in the face of outside pressures.

Irrigation allows for better and more diversified choices in cropping patterns and the cultivation of high-value crops. Successful irrigation is a crucial determinant of the world's future development because of its influence on the supply and price of food. As the debate over water policies sharpens, it is increasingly important that agricultural policy-makers help shape the nature of the debate and influence policy decisions.

THE HIDDEN TREASURE

The most prominent feature of the landscape of that garden is a large river that runs through it southwards. A part of its course is however hidden under the mountains and when updrawn with kindly thirst through the veins of porous earth it issues forth as a fresh fountain.

Rivers flow like God's blessings and the beauty and the fecundity not only of the garden of Eden but also of the entire earth are their gifts. As a feature of the surface of the earth a river is so prominent and imposing that there is no possibility of its being mistaken.

It is a surface flow in full view and therefore very familiar to us. It is always in our sight and can never be out of our mind. But there is another kind of rivers, so the early man thought. They flow underground out of our sight. Without them he found it impossible to account for the sources of water in wells, springs and fountains. These 'ingulfed' flows find a place in many ancient myths and legends. In real life also early man knew how to use naturally occurring springs and water holes as sources of fresh water. In fact the history of such use of groundwater goes to prehistoric times. During droughts and times of water scarcity early man must have deepened them. The ancient Iranians developed water-collecting galleries and tunnels, called 'kanats', which were several miles long and yielded considerable quantities of water. In course of time the people of Egypt and Afghanistan also learnt this primitive technology. The ancient Chinese perfected the art of drilling deep wells. With crude and primitive materials they built drilling machineries capable of drilling to the depths of 5000 ft, a feat not duplicated until modern times.

Though man developed the skill to exploit groundwater quite early he was yet to understand its nature. The ancient Greeks and Romans, for example, speculated about it. They thought that the limestone caverns common in their countries were passages which penetrated the entire earth to form underground rivers that were fed by the oceans whose salt water was freshened by a process of earth metabolism. As their countries were located in regions of relatively low precipitation, not the rains alone but also this water could only account for the water in their wells and springs. The ancient Indians went a step further. They believed that the natural regime of a river extended far beyond the visible surface of the earth. The Ganges, their most sacred river, not only watered their land but also the skies and the underground world where she was respectively called the Akash Ganga and the Patal Ganga. They had a myth also to explain the disappearance underground of another sacred river, Saraswati, in the dry sands of Rajputana and Sindh. It would not perhaps be too much to interpret these Indian myths as an intuition or intelligent guess about the existence of what is now called the hydrological cycle.

Sir Jagadish Chandra Bose, a modern Indian scientist, told an allegorical story in his Bhagirathir Utsa Sandhane (in search of the source of the Bhagirathi), which uses these myths and is in fact the story of this cycle. Till the beginning of the 18th century it was not recognized by educated people that rivers and groundwater were only two links in the vast and complex hydrological cycle. The credit for the final acceptance of this idea in modern times goes to two Frenchmen. Pierre Perrault (1608-80) showed that the total discharge of the Seine river from 1668 to 1670 was only one-sixth of the total volume of rainfall. Edme Mariotte (1620-84) verified Perrault's work by measuring the discharge of the Seine. He also measured the infiltration of rainwater into the subsurface grounds. Another noted French engineer, Henri Darcy (1803-58), was the first man to clearly state the mathematical law governing the flow of groundwater. The time when these Frenchmen were trying to understand and explore groundwater in good earnest is very significant. By their time the industrial revolution was about to make its appearance in the western European countries. Till that time the world population was small, predominantly agricultural and technologically backward and stagnant. Life was simple, its demands on the physical resources of the earth was not much. Everything was in plenty and seemed inexhaustible. This was particularly the case with fresh water. People were settled mainly near rivers and other surface water bodies which were easily accessible and were more than sufficient to meet their limited fresh water needs. This gave birth to an attitude of carelessness in their use of water. But now the population began to expand and settle in areas away from surface water sources.

Growing population, industrialization, urbanization, revolutionary changes in men's ways of life and their consumption pattern are now making ever increasing demands on all the physical resources of the earth. Most phenomenal has however been the growth of the demand on fresh water. Earlier it was needed mainly for agricultural and domestic purposes. Now it is needed for industries also. The traditional surface sources began to be exploited more extensively and intensively. The sources which so long seemed inexhaustible were found to be insufficient to meet the growing needs and had to be supplemented by supplies from other sources.

This is how intensive exploration and exploitation of groundwater began. Now it has become an essential part of fresh water supply, so much so that in many areas it is the only source of fresh water. Though stress on fresh water supply is being acutely felt in many areas of the world there has not been any significant change in man's attitude of carelessness. Greater human interference in the natural regimes of the surface water bodies has brought about their degradation. Similar has been the case with groundwater. Surface water is visible but groundwater is not. While we have been careless in our use of the visible resources it is no wonder that we should not be less careless in our use of this invisible resource. As it is out of our sight it stands very much in danger of being out of our mind. The degeneration of a river is apparent to everyone--the expert and the layman alike. But the threat to groundwater is not so. It needs special efforts to understand how precious is this hidden treasure and how criminally we have been squandering that treasure.

The importance of groundwater can hardly be overestimated. 97.5% of the total quantity of water contained in the hydrosphere is in the oceans and cannot be readily used because it is saline. The remaining 2.5% is fresh water and is essential for life. About 70% of this fresh water is in a frozen state in the Arctic, the Antarctic and the mountains, about 30% is under the ground and less than 1% is in lakes, reservoirs and rivers and streams. Thus in volume the position of groundwater is second only after the permanently frozen water and is about 30 times larger than all the water of rivers, lakes and other surface water bodies taken together. In arid and semi-arid countries it is the main, if not the only, source of fresh water, while in many water-rich countries it accounts for as much as 50% of the total supplies.

According to a European Environment Agency report, Groundwater Quality and Quantity in Europe, the share of groundwater needed by different countries of the continent to meet their total demand for fresh water ranges from 9% to 100%. About 80% of the rural water supplies in the USA and Canada comes from this source. In India, among other things, the intensive use of groundwater made the 'green revolution' possible. In highly industrialized countries about 50% of the water used for industrial purposes comes from groundwater sources. In some countries groundwater has begun to be used for unconventional purposes. In Canada, for example, it is being used as a source of energy. In that country the City of Moose Jaw has developed a geothermal heating system for a public swimming pool and recreational facility, the Carleton University of Ottawa uses groundwater to heat and cool its buildings and the health centre complex in Sussex, New Brunswick, has been utilizing an aquifer for thermal energy storage since 1995. More importantly, groundwater supplies about 30% of the water in rivers and streams. In developed countries about 70% to 80% of the population live in cities which depend mostly on groundwater for their water supplies. The same trend has overtaken the rest of the world. Today we have become so much dependent on groundwater that without it our survival is impossible. Therefore it has become very imperative for us to understand this precious treasure which nature has kept hidden underground.

WHAT IS THIS GROUNDWATER?

It is water that is found underground. But is it in some rivers or lakes as primitive man thought? Or is it, as Milton imagined, a river which passes a part of its course 'underneath ingulfed' in 'porous' earth and issues forth on the surface of the earth in the form of wells and springs 'with kindly thirst updrwan' i.e., by capillary action? Many such springs and wells have in fact given birth to rivers and streams, as it did in the garden of Eden. In certain respects Milton seems to anticipate what the scientists have found out, but his basic concept that it is a river is not correct.

According to the scientists it is neither a river nor is it collected in some underground lakes. It is not confined to only a few channels or depressions in the same way that surface water is concentrated in streams and lakes. Rather, it exists almost everywhere underground in the spaces between particles of rocks, gravels, sand and soil, or in crevices and cracks in rocks. Most of it comes from precipitation which infiltrates below the ground into the subsurface soil. Precipitation that falls on the surface of the earth is distributed in four main ways: some is returned to the atmosphere by evaporation, some is intercepted by vegetation and then evaporated from the surface of leaves, some percolates into the soil by infiltration and the rest flows on the surface of the earth as runoff giving rise to rivers, streams and lakes. Thus groundwater is an essential part of the hydrological cycle that involves the continuous circulation of water in the earth-atmosphere system.

In the upper layers of the ground this percolated water exists along with air and is called the aerated or unsaturated zone where the interstices or empty spaces are occupied partially by water and partially by air. The water found in this zone is called soil moisture. Below this zone occurs the saturated zone where all the interstices are filled with water. The topmost level of this zone is called the water table which rises or falls according as the quantity of water in the saturated zone increases or decreases as a result of heavy precipitation or drought or over-abstraction. The water table may be deep or shallow, it may be only a foot below the ground's surface or it may be hundreds of feet down. Some parts of the saturated zone contain more water than others because of their peculiar geological formation of permeable rocks or porous and loose materials like gravels, sand, limestone etc. These are called aquifers. They are like sponges which can absorb liquids that ooze out under pressure. They come in all sizes. They may be small, only a few hectares in area, or very large, underlying thousands of square kilometres of the earth's surface and extending across political boundaries. They may be only a few metres thick, or they may measure hundreds of metres from top to bottom.

Groundwater is not static but mobile, it is always in motion. But the rate of its movement varies according to the variable geological formation of the aquifers. As permeable or porous and loose or unconsolidated materials contain interconnected cracks or spaces that are numerous enough and large enough to allow water to move freely, in aquifers formed of such materials groundwater may move several metres in a day, whereas in aquifers formed of relatively impermeable or consolidated materials the rate of movement may be very slow, only a few centimetres in a century.

There are materials which are very porous yet impermeable. Clay, for instance, has many spaces between its grains, but the spaces are not large enough to permit free movement of water. Groundwater continues to descend beyond the saturated zone until at some depth it merges into a zone of dense rocks. Water is contained in the pores of these rocks, but the pores are not connected and water cannot migrate. Such water may be thousands of years old. Some aquifers are called fractured aquifers because they are rocks in which groundwater moves through cracks, joints or fractures in otherwise solid rocks. Examples of such aquifers include granite and basalt.

Limestones are often fractured aquifers, but here the cracks and fractures may be enlarged by solution, forming large channels or even caverns. Limestone terrain where solution has been very active is called karst. Sandstone is a porous media, but by solution it may become so highly cemented or re-crystallized that all its original spaces are filled and as a result it no longer remains porous. If however it contains cracks it can still act as a fractured aquifer. Aquifers may again be either unconfined-those that are bounded by the water table-or confined-those that lie beneath layers of impermeable materials. The aquifers of the latter kind are sometimes called artesian wells. The water in these wells rises higher than the top of the aquifer because of confining prèssure. If the water level rises above the ground surface a flowing artesian well will occur. Groundwater usually flows downhill with the slope of the water table. Like surface water it flows toward, and eventually drains into streams, rivers, lakes and the oceans. In aquifers underlying surface drainage basins this flow does not always mirror the surface flow. Groundwater may move below the ground in directions different from those of the water flowing overland.

Groundwater moves mainly because of the operation of the hydrological cycle. The aquifers receive their supplies from the percolation of water from precipitation. This is called their charging. Eventually this water reappears above the ground by flowing into rivers, streams, lakes and the oceans. This is called discharge. The extent of charging of the aquifers depends on precipitation, the slope, soil type and the vegetable cover of the surface ground above them and the geological formation of the aquifers themselves and their consequent capability to absorb water. This also determines the quantum of the surface flow called runoff. If for any reason the underground aquifers are less capable to absorb enough water from precipitation the surface flow will be more and may cause floods. In course of time nature establishes a kind of balance among the various links in the hydrological cycle. As an integral part of this cycle groundwater also develops a regime of its own. At times this balance and the regime may get disturbed by geological and cosmic extreme events like earthquakes, droughts, heavy rains and abrupt shifts in the climatic cycle. But nature, when left to itself, has the capacity to restore them or to re-establish a new balance and regime.

Since the industrial revolution men are exploiting groundwater at an ever increasing rate to meet their ever increasing fresh water needs and thereby interfering too much in its natural regime. Now in many areas of the world the rate of withdrawal of groundwater from aquifers is much in excess of the rate of their recharging. Such over-exploitation of groundwater has serious consequences. Continued over-extraction of groundwater will cause the water table to fall resulting in the ultimate depletion of aquifers. Renewal of water in rivers, for example, takes only a few days whereas in case of groundwater it takes 1400 years. Thus if an aquifer gets depleted due to overdrawl of its water in excess of the rate of its recharging it will need these many years to get replenished. And what will happen in the mean time to the vast population which depends on this source for its survival and the material civilization it has built upon it? That civilization will come crashing like a house of cards and that population will be extinct in a matter of days.

Other adverse effects of such over exploitation of groundwater are deprivation of rivers, streams and lakes of an important source of their supplies, subsidence or caving in of land above the aquifers, drying up of wetlands and other small water bodies disastrously affecting the ecosystems they support and salt water intrusion in coastal regions rendering them unusable. Many areas of the world are already experiencing these adverse effects and in days to come many more areas will follow suit. Due to over-drafting of groundwater parts of Mexico City, for instance, have subsided as much as 10 metres in the past 70 years, resulting in a host of problems in its water supply and sewerage systems. According to the US Geological Survey in the United States more than 17,000 square miles have been directly affected by subsidence.

Over-abstraction may give rise to another danger. An aquifer which is used by more than one country or region may become a bone of contention amongst those countries or regions. Excessive irrigation on the other hand may cause water table to rise and lead to water logging making the land saline. We often ignore the geological aspect of groundwater and the adverse effects that its over-abstraction may have on the underground geological structure of an area. Engineers must consider groundwater when planning almost any kind of structure, either below or above the ground. Ignoring the slope stability can be costly and dangerous. Groundwater is a major force in geological changes. The fluid pressure exerted by groundwater, for example, play an important role in the occurrence of earthquakes. Geologists also know that the movement of water through underground geologic formations controls the migration and the accumulation of petroleum and the formation of some ore deposits.

We are threatening groundwater not only quantitatively but also qualitatively. Water is never found in a pure state in nature and both groundwater and surface water may contain many constituents which make their use without appropriate treatment unfit for specific purposes. The suitability of water for a given use depends on many factors and considerations important for surface water are equally applicable to groundwater. But one very important difference between the two is that as groundwater flows through an aquifer it is naturally filtered. This filtering, combined with the long residence underground, makes groundwater usually free from disease-causing micro-organisms. Natural filtering also means that groundwater usually contains less suspended material and un-dissolved solids than surface water. This is the reason why it is generally assumed that it is safe and is therefore used mostly without any treatment. But it may not be so. The rain water which charged the aquifer itself might have been polluted by the polluted atmosphere. The acid rain is an example.

Again, one of the most important natural changes in groundwater chemistry occurs in the soil and in course of its journey through the aquifer it dissolves certain substances it encounters. Who knows if arsenic, the greatest menace now found in groundwater throughout the world, is due to this? Most dangerous are the innumerable ways in which we are contaminating groundwater by our various activities and making it unsafe. There is a popular misconception that as it is not exposed like surface water it is less vulnerable to pollution.

Human activities which are mainly responsible for pollution of groundwater are those related to agriculture and industry. Modern agriculture is very intensive and uses heavy doses of chemical fertilizers, pesticides and herbicides containing highly toxic compounds. Industrial processes also involve the use of many toxic materials. Through our agricultural and industrial activities we also generate huge amounts of hazardous wastes both liquid and solid. They pollute not only the soil, air and surface water but also find their way to the underground aquifers. Other factors responsible for groundwater pollution are leaking pipelines and storage tanks containing petroleum products, septic systems, municipal landfills, livestock wastes, leaky sewer lines, wells for disposal of liquid industrial wastes, toxic substances from mining sites, dumping grounds for disposal of solid and liquid wastes, fly ash from coal-fired power plants, contaminants in rain water etc.

In cold countries road salt is a major source of contamination. Groundwater pollution is a serious problem and is very costly and difficult, if not impossible, to solve. That is so because of its relative inaccessibility, huge volume and slow flow rate. Once pollutants enter an aquifer, the environmental damage can be severe and long lasting because of the very long time needed to flush out the pollutants, if it is at all possible to do so. Contamination of groundwater is also very insidious, it takes many years to show up in water withdrawn from wells. By the time it is detected, it may be too late to prevent serious contamination. And as a major link in the hydrological cycle polluted groundwater pollutes surface water also.

Thus it is obvious that though the groundwater reserve is huge we cannot use it as much as and in any manner we like. Because of the threats to groundwater and its importance not only for the maintenance of the present standard of our material civilization and a sustainable development, but also for our very survival it is urgently necessary to stop its overdrawal and pollution. There should be continuous monitoring to detect and prevent them. Withdrawal of water from aquifers found polluted should be immediately stopped and appropriate steps should be taken for its early purification. It is a very difficult task for various reasons.

It is a highly technical job and there is a great shortage of groundwater experts throughout the world. Costs involved are also huge. Even in a rich country like the U.S.A. many states are unable to bear the burden of all the functions involved and some of it, like the UIC or Underground Injection Control Program, is done for them by the federal government.

Exploration of groundwater started much later than surface water and sufficient and reliable data are therefore not often available. Finally, the root cause of excessive demands not only on groundwater but also on all the physical resources of the earth is the abnormal growth of population. Unless that cause is removed it is impossible to solve the problems of groundwater. And unfortunately the rate of population growth is comparatively much higher in poorer countries. To cap it all, many of these poor countries happen to be poor in fresh water resources also. According to the UNEP Report, Global Environment Outlook-2000, India is one of those countries which are currently withdrawing 20 to 40% and will be withdrawing around 40% of their water resources by the year 2025 AD when they will face a severe fresh water crisis. According to the Report of the World Commission on Dams in India the groundwater table is falling at an alarming rate of one metre per year. It is therefore needless to say that we have a special reason to be careful in the use of our water resources. To save ourselves we have to save our groundwater too.

3

Overexploitation and Groundwater Mining

The limits how much water can be extracted from a finite groundwater aquifer are economic and environmental. When water is pumped out faster than it is recharged by the natural processes, the water level in the aquifer drops, and the distance the water must be raised to the surface increases. Eventually, either the energy costs rise to the point that exceeds the value of the water, or the water quality falls bellow acceptable levels. At this point pumping must cease.

Although many aspects of what can be called overexploitation are not new in Hydrogeology, this concept is still poorly defined and subjected to varying interpretations by the different kinds of specialists, managers, policy makers and the public. A set of terms, not always fully equivalent are in use for concepts related to overexploitation, such as overpumping, overdraft, overdevelopment and groundwater mining.

The debates between experts and decision-makers on the management of groundwater resources, as regards the overexploitation of aquifers, are based on two main questions: (a)How should we assess whether an aquifer is being overexploited, or predict if this may happen as a result of planned new exploitation? According to what criteria?; and (b) Is the overexploitation of an aquifer always undesirable, "bad management" that should be prohibited and prevented, or is it permissible or even advantageous under certain situations, and what impact should be anticipated and compensated for?

It is argued in this paper that non-equilibrium or an unbalanced regime of an aquifer can not simply be identified as overexploitation. The assessment of overexploitation is relative to the criteria used, which are themselves linked to the resource-management objectives.

These criteria can therefore be:

(a) *purely physical and quantitative*: depletion of resource, non-equilibrium etc.;

(b) *qualitative*: degradation of water quality;

(c) *economic*: non competitiveness, or more broadly-the all direct costs and external costs greater than collective advantages;

(d) *social*: conflicts of use between unequal developers with detrimental effect suffered by third parties;

(e) *environmental*: damage to the natural environment, especially to sensitive aquatic ecosystems.

To evaluate possible aquifer overexploitation, not only detrimental (negative) effects have to be considered, but also beneficial (positive) ones. Otherwise a biased assessment may be reached. In some extreme cases of severe water shortage beneficial aspects may dominate over detrimental ones.

Various aspects and approaches in groundwater management of heavily overexploited aquifers are further analysed in the paper, as well as possible alternatives to it, like artificial recharge, reclamation and reuse of wastewater, desalinization, weather modification and finally demand modification. Key-words: Groundwater Management, Groundwater overexploitation, Groundwater mining

INTRODUCTION

Since the publication of the Bruntdland Report "Our Common Future" in 1987, sustainable development has been adopted as governing philosophy (WCED, 1987). Despite the broad acceptance of this philosophy and its recognition on various important meetings concerning Integrated Water Resources Management in New Delhi(UNDP, 1990), Delft

(Alaerts et al., 1991), Dublin (ICWE, 1992),and Delft (Hall et al, 1993), we are witnessing water management practices overstepping ecological limits. Aquifer overexploitation and groundwater mining are reported from all parts of the world.

The limits how much water can be extracted from a finite groundwater aquifer are economic and environmental (Gleick, 1993). When water is pumped out faster than it is recharged by the natural processes, the water level in the aquifer drops, and the distance the water must be raised to the surface increases. Eventually, either the energy costs rise to the point that exceeds the value of the water, or the water quality falls bellow acceptable levels. At this point pumping must cease (Postel, 1993)

Although many aspects of what can be called overexploitation are not new in Hydrogeology, this concept is still poorly defined and subjected to varying interpretations by the different kinds of specialists, managers, policy makers and the public. A set of terms, not always fully equivalent are in use for concepts related to overexploitation, such as overpumping, overdraft, overdevelopment and groundwater mining.

The debates between experts and decision-makers on the management of groundwater resources, as regards the overexploitation of aquifers, are based on two main questions (Margat, 1991):-How should we assess whether an aquifer is being overexploited, or predict If this may happen as a result of planned new exploitation?

According to what criteria?; and-Is the overexploitation of an aquifer always undesirable, "bad management" that should be prohibited and prevented, or is it permissible or even advantageous under certain situations, and what impact should be anticipated and compensated for?

Young (1992) emphasizes that aquifer management, especially of overexploited ones, which deals with a complex interaction between human society and the physical environment, presents an extremely difficult problem of policy design. Establishment of a groundwater management system involves selection among strategies, which variously satisfy competing objectives and criteria. The most suitable policy will include a mix of structural, institutional and cognitive measures.

SAFE YIELD, AQUIFER OVEREXPLOITATION AND GROUNDWATER MINING

Hydrogeologists and groundwater managers are concerned about the sustainable economic output when there is excessive groundwater withdrawal. Together with engineers they developed the concept of safe yield. Conversely, overexploitation is largely point of view referring to the consequences of intensive groundwater use, as perceived by environmentalists, sociologists, news media and public in general, and places more emphasis on the adverse or detrimental aspects, sometimes even with apocalyptic undertones (Custodio, 1992).

To evaluate a situation that can be termed overexploitation, not only hydrological aspects have to be taken into account, but also economic, social and political ones, as well as the point of view of the stakeholders and all persons involved (groundwater exploiters, water administrators, water managers, land-use planners, economists, local people and environmentalists).

Overexploitation is term oftenly used when the rate of abstraction exceeds the so-called 'safe yield' (or sustainable yield) with the formation of overdraft areas. In another interpretation, undesirable results occur when the groundwater storage cannot be replenished by a natural recharge in a reasonable period of time, or when prolonged abstraction results in the intrusion of the saline water, or in deterioration of water quality (Balek, 1989).

The term groundwater mining is used when conscious and planned abstraction rate greatly exceeds aquifer recharge (UN, 1992).

In some arid and desert areas there are enormous quantities of groundwater stored in aquifers. Present day studies based on tracing with different isotopic (radiometric) elements and computer model calculation of groundwater flow velocities and age, indicate that these waters infiltrated long before present day. They are often described as 'fossil waters'. Exploitation of these waters, as well as connate and juvenile waters, considering that the replenishment is negligible or non-existent, is analogous to that of any other non-renewable mineral resource, and for this reason is referred to as groundwater mining.

Safe Yield

Safe yield is defined as the amount of water, which can be taken from the aquifer indefinitely without producing an undesirable result. From a hydrological point of view, the maximum safe yield is equal to the long-term mean annual recharge of any origin or, in other words, to the amount of water which under normal circumstances leaves the basin as a natural baseflow. We can recognise (Balek,1989):

- potentially exploitable groundwater resources which represent a maximum close to the live storage, and are limited only by technical problem of setting an adequate system of boreholes to utilize the groundwater to the maximum; and
- actually exploitable groundwater resources governed by technical, environmental and legal requirements on the minimum baseflow and/or minimum groundwater level.

Safe yield, called also sustainable yield, is considered as the upper limit of exploitation. But Walton (1970) concluded: "From the water users' point of view, the so called sustained exploitation means that the water supply has being frozen at a certain level because of the seemingly unfounded apprehensions of a hydrologist"

Safe yield is determined using water balance equations for a representative period of time. The question to be answered is, to what extent can the groundwater resources be exploited during dry years? This problem is closely related to the probability of occurrence of low river discharges, because during prolonged dry period a great part of the river discharge, if not all, is supplied from groundwater resources. Statistical methods can be applied to determine the safe yield from monitored groundwater table level data or rainfall data, provided that correlation exists.

Aquifer Overexploitation

Effects of aquifer exploitation, do not only depend on the volume being abstracted, but also on the distribution of withdrawals and the well pattern (Van der Gun et al, 1992). Thus, the safe yield of an aquifer is a complex function that changes with time. It cannot be considered as a fixed value for an aquifer. The problem is complicated with changes in natural recharge due to land-use modifications and different forms of artificial recharge (Custodio, 1986). Margat (1992) argues that non-equilibrium or an unbalanced regime of an aquifer can not simply be identified as overexploitation. He proposes that evaluation of overexploitation depends on the agreement on the meaning of undesirable results or adverse situations caused by it. The assessment of overexploitation is relative to the criteria used, which are themselves linked to the resource-management objectives. The criteria will not be the same whether the objectives are:

- maintaining conservative conditions for exploitation at a given stage of development;
- searching for ways to develop the use of resource by intensifying exploitation (either long-term, and therefore limiting the use to renewable resources, or by temporarily withdrawing part of the non-renewable resource); or
- minimising detrimental impacts on the users of surface waters or the occupants of the land.

These criteria can therefore be:

- *purely physical and quantitative*: depletion of resource, non-equilibrium etc.;
- *qualitative*: degradation of water quality;
- *economic*: non competitivnes, or more broadly-the all direct costs and external costs greater than collective advantages;
- *social*: conflicts of use between unequal developers with detrimental effect suffered by third parties;
- *environmental*: damage to the natural environment, especially to sensitive aquatic ecosystems.

The diagnosis of overexploitation a priori, and especially forecasting it, presupposes independent and fairly reliable knowledge of both the average recharge and the total abstraction that have to be compared, which is often far from practicable with the necessary accuracy.

For different reasons, Margat (1992) considers the uncertainty in the assessment as high on both sides and drawing up a 'balance' as a way of demonstrating nonequilibrium unless it is very obvious. In practice, overexploitation is diagnosed a posteriori, on the basis of the observed manifestations of the prolonged unbalanced regime: continuous falls of the water level and possible consequences on the boundary of the aquifer system and the quality of the water.

However, model simulations can be carried out to predict and study the long-term impacts of aquifer overexploitation and their extents (Van der Gun et al, 1992). By comparing different pumping regimes and/or relocation of wells, models can contribute to choosing the most appropriate groundwater management strategies in an intensively exploited aquifers.

To evaluate possible aquifer overexploitation, not only detrimental (negative) effects have to be considered, but also beneficial (positive) ones. Otherwise a biased assessment may be reached. In some extreme cases of severe water shortage beneficial aspects may dominate over detrimental ones.

Groundwater Mining

When there is continuous water table lowering due to groundwater abstraction greatly exceeding recharge, aquifer reserves are being used in a process similar to mineral mining. Once the aquifer is depleted, especially in arid areas, it will take generations or even centuries to recover.

Since fresh groundwater in the aquifer is not beneficial to anyone, except to sustain some environmental conditions, in principle there is no objection to 'mine' the reserves for beneficial use if environmental values and the rights of future generations are duly considered.

If it is well known that reserves are limited in quantity (or time), to be confident that such a policy is not against the general interest, an economic study has to be undertaken to determine:

- the interest rate during period of repayment of capital investment;
- local benefits;
- general interests satisfied by development;
- the foreseeable consequences, including ecological impacts;
- equity consideration with respect to future generations.

In coastal areas groundwater mining can be undertaken by pumping from the aquifer until it is completely encroached with sea water. It can be reasonable decision if it is carefully planned for development.

Groundwater mining is well known in North Africa (Lybia's "Great Man-Made-River Project" mines fossil water from Sahara Desert and pumps it near the Mediteranean Sea). Also in the Arabian Peninsula there are cases of both overexploitation and groundwater mining. Well planned 'groundwater mining' is not to be considered 'overexploitation' if things go as planned, even if the effects are the same. It can be termed overexploitation when there are serious deviations from what was planned or when there is no planning at all.

Undesirable Effects of Overexploitation

Changes following aquifer overexploitation that can be termed undesirable results, as stated in most of the consulted literature cam be summarized as:

- groundwater level deepening;
- spring and river flow diminution or/and wetland surface reduction;
- degradation of groundwater quality, either salinity increase or the increase of certain undesirable constituents;
- land surface changes in the form of generalized or local land subsidence, or ground collapse.

These changes produce a series of undesirable results with the following aspects (Custodio, 1991):

- *hydrological aspects*: groundwater decrease reduces transmissivity in water table aquifers and the yield and specific yield of the wells. This is more acute in thin heterogenous aquifers and in fractured rock aquifers. Well and water galleries have to be deepened or redrilled. Pumping machinery becomes inadequate, works outside the performance range or have to be replaced. Their energy supply facilities have to be enlarged or substituted. There is a decrease in spring and river flow and uptakings reduce their yield or decrease the flow frequency distribution. The entire water resources system is changed;
- *water quality aspects*: wells and water galleries have to be abandoned due to deteriorating water quality, and some treatment has to be introduced before use, including in some cases desalination facilities. The repeated use of low quality groundwater may seriously damage agricultural soil, vegetation, animal raising and industrial production, and in some cases may be the cause of health problems. The coastal aquifers deterioration may appear as the result of the formation of saline upconings;
- *economic aspects*: exploitation becomes more costly energywise and new investments are needed for well and gallery deepening and substitution, for pumping machinery, and energy facilities, and for new water transport lines.
- Eventually, pumping must cease because the pumping costs exceed the value of the water pumped or the aquifer is already dry;
- *environmental aspects*: water table reduction in shallow water table aquifers may affect vegetation and agriculture, killing phreatotypes and plants able to use capillary water. This is especially noticeable in wetlands and low-lying small coral islands. The fauna associated

with affected vegetation is equally damaged and many migratory animals may disappear from the area;

- *morphological and geotechnical aspects*: this mainly related to undesired regional or local land subsidence in unconsolidated sediments and/or ground sinking by collapse in shallow carbonate or gypsum formations. The results are damages to structures, buildings, basements, roads, railways, canals, etc., and increased risk of inundation after heavy rains, storm surges and high tide. Also, abandonment of wells in low-lying, intensively exploited areas may produce inundation and uplift of constructions made when water table was artificially lowered. This is undesirable after exploitation effect;
- *legal aspects*: water rights and pumping capacity of other abstractors, public or private, may be affected. Depending how existing water rights are considered, a normal hydrodynamic evolution may be termed overexploitation and thus, under given legal prescriptions, lead to introduction of restrictions. A side effect is the extraordinary increase of complaints and legal litigation.
- *social aspects*: the increased costs of obtaining water, water quality deterioration and water availability reduction may create social unrest and accelerate the loss of employment. But existing experience is not so dramatic. In many cases, there is an adaptation, easy or painful, and no catastrophic situations are actually known, in spite of predictions in heavily overexploited areas (Custodio, 1989). Nevertheless, non-renewable groundwater resources consumption today is a loss of opportunities for the coming generations. It is an economical and ethical issue and should be carefully considered.

Perhaps, the most detrimental cause of overexploitation is the ignorance of what is happening and the negligence in getting the needed data to correctly evaluate the hydrogeological and economical situation. In the long run, a balance will somehow be reached or the overexploitation will finally be abandoned, but meanwhile a major investment in money, effort, and in hope, will be lost (Foster, 1991).

Another detrimental effect of overexploitation, is the irresponsible overreacting of water authorities, especially when they are poorly informed or lack the scientific technical skills to evaluate the physical problem. To try to protect existing wells, the decision to avoid further water level decline usually leads to interdiction of drilling new wells there or nearer the aquifer recharge area. This decision may prove to defend existing interest, but the aquifer may be underexploited, losing large quantities of water since the first by law have installations that do not allow correct exploitation of aquifer resources (Custodio, 1992).

Benefits of Groundwater Exploitation

Beneficial aspects of groundwater exploitation, mainly from the economic point of view, can be summarized as follows:

- development in simple stages, drilling new wells only as the need for additional water arises. This allows fro a better fit between demand and supply, progressive rather than costly intensive financing and the payment of works from revenues of the part of the project already finished.
- little surface area is needed for wells and related water winning facilities, new surface area is more easily acquired, there is no displacement of towns, agricultural lands or transportation ways, and rivers are much less affected.
- local manpower can be used, maintenance cost are low and may be solved locally. In remote rural areas pump maintenance and energy supply may be a major problem, but when investment costs and risks of failure are considered, groundwater is still most suited if the savings and benefits are used to create maintenance and energy services. They contribute to the development of the area, as well.

- water resources availability is increased by evaporation reduction. Evaporation may be enormous in shallow surface water reservoirs, sometimes leaving a brackish residuum.
- besides, the lowering of groundwater levels allow for valuable utilization of shallow groundwater, which under natural conditions, would be lost through evaporation. Abstraction in an alluvial aquifer creates a storage capacity for flood water to infiltrate.
- groundwater is in most instances suitable for human consumption and safe without or with only simple treatment is wells are properly constructed and maintained. wells may produce water close to where it is needed, thus saving costly-to-install and to-maintain water lines, the aquifer playing this role, at least to some extent.
- groundwater is generally more efficiently used in the rural world since it is close to the user. Also complex and sometimes unfair subsidizing can be avoided or reduced, and administration can be made easier.

These benefits are not absolute and have to be weighed against other alternatives, case by case taking into account local physical, human, economic, environmental and legal circumstances. Benefits exist even if exploitation leads to some kind of overexploitation. What is needed is to know the consequences and extract a part of the benefits to compensate for them, now, or in the future, when the investments begin to produce the expected economic return.

In most instances groundwater allows the temporary solution of acute problems. Especially when awaiting for better but difficult, long-term, expensive and uncertain alternatives on the basis of presumed aquifer overexploitation. The possibility to avoid megaprojects that can cripple the economy of a developing country (generally financed by other countries or international Institutions) also has to be considered. However, aquifer, than, have to be studied and monitored, undesirable effects corrected to the maximum, and provisions have to be made to cope with reserve depletion and ecologic impacts. Overexploitation is an acceptable policy only if planned with specific aims and as long as negative consequences have been technically evaluated by decision makers and they are economically and socially acceptable. This is the case if groundwater mining is used to induce a cycle of economic development, which will give way to substitution of more expensive water at a later date or to new technology improving the water use. Even so, the issue of intergenerational equity can arise and it has to be considered with due seriousness.

Management of Aquifers' Utilization

Aquifer management, which deals with a complex interaction between human society and the physical environment, presents an extremely difficult problem of policy design.

If we suppose that aquifer were managed according to the economist's preferred criterion of maximum net return to the entire aquifer, than the rate of abstraction would be the one that gives the maximum Present Net Benefit. The rate obtained would be much lower than the rate at open access equilibrium, thus preserving the aquifer on a sustainable basis and generating a much larger present value of economic return. Economist would then define 'overexploitation' as any pumping rate in excess of that which yields the maximum present value of net benefit (Young, 1991). If externalities are included in the analysis, represented by 'damage functions', the result for an optimal level of abstraction is similar, but even more restricted, yielding even smaller rate of abstraction.

A successful method of managing aquifer exploitation, as a part of regional water resources management, should have the capability of balancing competing demands so that actual operating policies optimize the net benefit to the region (Buras and Nunn, 1989).

Aquifer overexploitation may be caused by either, or both, of two classic types of social dilemma problems (Young, 1991):

- aquifers are typically common pool resources, in which a migratory subtractable resource is exploited under an unrestricted rule of capture. Those using the resource are little motivated to pressure its value, and the collective inefficiency of a pumping race is likely to result;

- extensive exploitation of aquifers often imposes unwanted damages to third parties-external costs or "externalities". External costs are unwanted and uncompensated costs imposed on third parties who do not themselves benefit from the exploitation activity. According to the exploitation strategies, three methods of use (of management) of the reserve of an aquifer are therefore conceivable and practicable (Margat & Saad, 1983: Margat, 1991):
- a strategy of maximum and lasting exploitation of the renewable resources, in a regime of dynamic equilibrium, with average abstraction greater than average recharge (possibly enhanced by boundary effects), without taking into account seasonal or even possible annual variations. Thus, after a decrease in the initial phase of nonequilibrium, the stabilized reserve is used, usually as a regulatory factor (annual, multiannual)-its natural regulatory function being amplified-under a constraint of preserving a minimum flow rate at the outflow boundaries of the aquifer or preserving the freshwater/ seawater equilibrium in a coastal aquifer.
- a strategy of repeated exploitation of the storage in a prolonged unbalanced regime, that may be intentional or unintentional in the initial phase. Than, abstraction increasing or stabilized) is greater than recharge (even enhanced by boundary effects). A second phase involves reducing abstraction to restore the equilibrium. In this case, the depletion of the reserve contributes largely, sometimes predominantly, to the production of the water. This phase of medium, to long-term, duration is limited either by external constraints or by reduction of productivity of the wells. In second phase of possible re-equilibration, the reserve may be either:

(a) stabilized, on average, bringing abstraction close to recharge; or

(b) in part restored, by reducing abstraction bellow recharge (repayment of the loan") and sometimes by artificially increasing recharge, and than stabilising, on average, in this new condition.

- a strategy of mining or exhaustion exploitation, with abstraction from the start of operation much greater than the average recharge. In this case depletion of the reserve provides most of the produced water. The exploitation is in long term more or less limited when drawdown becomes excessive without later returning to a regime of re-equilibration. The recovery of the reserve may be to slow and sometimes hindered by irreversible degradation of the storage capacity of the reservoir due to subsidence.

The diversity of hydrogeological conditions and types of aquifers, of rates of recharge and of volumes of recharge, present a large range of possible aquifer developments. Several criteria are appropriate for assessing management strategies (Young, 1991):

- economic efficiency: most value of output from any given input, ability tomaximize net economic returns and in long run dynamic, flexible system, responsive to the changing conditions;
- equity;
- security: satisfaction of minimum human needs;
- liberty: ability to act freely as long as it doesn't interfere with others;
- avoiding harm;

Secondary criteria are:

- local control and popular participation;
- orderly conflict resolution processes;
- information intensity;
- ease of monitoring and enforcement;

- social considerations;

We can distinguish two basically different approaches in management of overexploited aquifers: structural and nonstructural.

Structural approaches generally involve developing some alternative source of water supply, whether conventional (reservoir/conveyance schemes, conjunctive use of surface and groundwater) or unconventional (desalination, weather modification through cloud seeding etc.). Nonstructural approaches are policies of demand management. They include: (a) cognitive methods to modify human behaviour; (b) institutional arrangements to coordinate activities of individual water users; (c) administrative organizations as a necessary element of institutional arrangements. Incentive and sanction system can influence individual pumper behaviour. Two general types of incentive-based strategies can be employed to influence the behaviour of pumpers in an overexploited aquifer(Young, 1991):

- financial incentives, both positive and negative:
  (a) pumping charges or taxes, subsidies;
  (b) quantity-control approaches: permits, pumping quotas, transferable pumpingentitlements, use of water-rights markets;
- monitoring and enforcing pumping controls.

Aquifer exploitation is performed by humans and understanding how people respond to hydrologic facts, economic opportunities, and institutional constraints is important in designing appropriate groundwater management policies. Diagnosis and successful resolution of aquifer overexploitation have to be achieved by a coordinated interdisciplinary approach.

ALTERNATIVES TO GROUNDWATER MINING

Artificial Recharge of Aquifers

When natural recharge is smaller than the pumping rate, the artificial recharge of the aquifer can be undertaken using one or another of the techniques currently available. A main point of these operations is temporary storage within aquifer of surface water not currently delivered for use, originating from a number of possible sources. Examples of these sources are waters from other hydrological units, enhancement of infiltration of flood waters in ephemeral streams, and treated effluent from waste water treatment plants.

It must be noted that only unconfined (phreatic) aquifers are suitable for artificial recharge. Additional problem in subsurface storage is the change of water quality: change, usually increase, of total dissolved solids (TDS), questionable effectiveness in filtration by the soil mantle overlying the aquifer, especially in the long run.

Artificial recharge and conjunctive use of surface and groundwaters is probably the one of best solutions to cope with ever increasing demand for fresh water...

Reclamation and Reuse of Wastewater

Treated municipal wastewater, provided that it is treated adequately, can be a possible irrigation water source. A distinction is made between industrial and municipal wastewater. From a regional point of view, the integration of reclaimed waste water within the water resource system raises number of inter-related organizational, managerial, engineering and institutional problems.

Desalination

Ninety-seven percent of all water on the planet earth is in the oceans, unsuitable for human consumption or irrigation use. This has lead to great interest in devising ways of removing salt from water in hopes of providing unlimited supplies of fresh water. Despite the lack of technical obstacles to desalination, the high energy cost of these processes continues to make it an elusive goal.

In the energy-rich arid and semi-arid regions of the world with a great discrepancy between water demand and supply, desalination is increasingly important option. For poorer countries desalination continues to be too expensive to pursue on a large scale (Gleick, 1993). The disposal of brines in inland plants can also be a problem (Buras & Nunn, 1989).

Weather Modification

Cloud seeding technology has developed over the last 35 years as a means of augmenting precipitation. The possible augmentation is fractional, and where successful, is in the range of 5-20 %. However, even these small increases in precipitation may generate important economic benefits. The planning, implementation and management of weather modification (cloud seeding) requires an interdisciplinary team of scientists and professionals.

Demand Modification

In recent years demand management is recognised as one of the most important instruments that water resources managers have to develop and use during the coming years (UNDTC, 1991; ICWE, 1992). Demand oriented measures include an array of technical, administrative, legal, political and operational instruments (Savenije, 1995).

Alternatives to groundwater overexploitation are neither simple nor easy. Those mentioned are the major ones, yet not exclusive. The applicability, especially in regions of severe water scarcity is often complicated by lack of financial, technical organizational framework.

Conclusions

Non-equilibrium or an unbalanced regime of an aquifer can not simply be identified as overexploitation. The assessment of overexploitation is relative to the criteria used, which are themselves linked to the resource-management objectives. These criteria can therefore be: (a) purely physical and quantitative: depletion of resource, non-equilibrium etc.; (b) qualitative: degradation of water quality; (c) economic: non competitiveness, or more broadly-the all direct costs and external costs greater than collective advantages; (d) social: conflicts of use between unequal developers with detrimental effect suffered by third parties; (e) environmental: damage to the natural environment, especially to sensitive aquatic ecosystems.

To evaluate possible aquifer overexploitation, not only detrimental (negative) effects have to be considered, but also beneficial (positive) ones. Otherwise a biased assessment may be reached. In some extreme cases of severe water shortage beneficial aspects may dominate over detrimental ones.

Overexploitation may even be an acceptable policy if planned with specific aims and as long as negative consequences have been technically evaluated by decision makers and they are economically and socially acceptable.

Depending on the exploitation strategies, various methods of use (of management) of the reserve of an aquifer are conceivable and practicable, and they should be assessed for various criteria. We can distinguish two basically different approaches in management of overexploited aquifers: structural (conventional and non-conventional alternatives to overexploitation and mining) and non-structural, comprising policies of demand management.

4

Groundwater Development in India

At the end of the 20th century, the population of India crossed one billion mark. India has mainly an agricultural economy where land holdings are generally small. Due to steep rise in population, the demand of water has increased considerably over the last 50 years resulting in overdraft and water quality deterioration.

The paper discusses major aspects of groundwater development and connected socio-economic issues.

INTRODUCTION

In India about 70% people live in rural areas. However, in recent years there is a greater trend of migration of the rural population to urban areas. Due to steep increase in population along with associated activities, water is becoming a scarce resource coupled with quality deterioration.. It is estimated that the entire utilizeable water resources of the country would be required for various uses by the year 2025 A.D.

The maximum use of water is for irrigation. Surface water is Government owned but groundwater is private i.e. it belongs to the land owner.

There has been an indiscriminate use of water as there is no control on its use either by the Government or public organizations. Recently attempts are being made for an equitable distribution of this scarce resource for the economic development and well being of the society.

GROUNDWATER OCCURRENCE

The vast expanse of alluvial formations in north India deposited by the Ganga-Brahamputra river system, is most productive. It consists of thick pile of sand, silt and clay with occasional beds of gravel, extending to a depth of about one km. Tubewell irrigation is being practiced in this area from the last more than one hundred years. The area is bestowed with good rainfall and hence good recharge.

The sedimentary formulations comprising of sandstone, shale and limestone, varying in age from Pre-Cambrian to Tertiary, occur as isolated pockets in different parts of the country. Permeability of older formations is quite low and therefore well yields are poor. However, rocks of younger ages (Mesozic and Tertiary) are more productive forming artesian aquifers at places.

The main volcanic suite of rocks is represented by Deccan Traps occupying an area of about 500, 000 km^2 in the western and central parts of the country. A number of basaltic flow units are identified of Upper Cretaceous to Lower Eocene age. Part of the Deccan Trap area has semi-arid climate. Low rainfall and rugged topography results in poor recharge to groundwater.

Crystalline (igneous and metamorphic) rocks of Pre-Cambrian age occupy nearly two-third part of the country. The permeability is poor except in weathered and fractured parts. Groundwater from weathered horizons is tapped by large diameter shallow (30-25m) wells. Small diameter (4 to 6 cm) bore wells of depth 30 to 70 m tap deeper fractured zones. Geophysical and remote sensing techniques have greatly increased the success rate of well drilling. Natural recharge from rainfall is being supplemented by artificial recharge.

WATER RESOURCES AVAILABILITY AND DEMAND

The average annual rainfall in India is about 1170 mm depth with a total volume of 4000 km^3..However, its distribution is greatly uneven in space varying from about 100 mm in the western parts to more than 1100 mm in the north-eastern parts of the country. The total utilizeable water potential is 1140 km3, out of which 690 km3 is from surface water and 450 km3 from groundwater. Water utilization for the year 2000 and the projected demands for the year 2025 are given in Table below.

Table: Estimates of Annual Water Demand in India (in Km³)

1. Purpose/Year	2000	2025
2. Domestic use	33	52
3. Industrial	30	120
4.Energy	27	71
5. Irrigation	630	770
6. Other uses	30	37
7. Total	750	1050
Distribution		
Surface water	500	700
Groundwater	250	350

Table above indicates that out of total water utilized in the country, 84% is used for irrigation, about 4.4% for drinking and municipal use, 4% for industry, 3.6% for energy development and the remaining 4% for other purposes. Table above also shows that nearly the entire utilizeable water resource of the country would be required to be put to use by the year 2025. This is mainly on account of increased demand of water for irrigation required to grow more food grains for the increasing population which is estimated to reach about 1.25 billion by the year 2025.

Groundwater Development

Out of the total replenishable groundwater which is estimated to be about 450 km^3,the present development is nearly 40%.However, its development has not been uniform. Groundwater contributes to about 50% of the total irrigated area of the country and 85% of the rural water supply is met from groundwater.

There has been a significant increase in the demand of groundwater for agriculture, domestic and industrial uses, which has caused over-exploitation of groundwater in some areas causing excessive lowering of water levels, seawater intrusion in coastal aquifers, and upconing of saline water in inland areas. In contrast to this, in canal irrigated areas, due to poor drainage and excessive irrigation, a rising trend of water level has taken place resulting in water logging and soil salinization. About 25,000 km^2 area under canal irrigation is affected by water logging.

The use of groundwater for irrigation has considerably increased from a meager 6.5 km^3 in 1951 to over 50 km^3 in 1997.The number of wells has also considerably increased from merely 4 million in 1951 to nearly 17 million in 1997. The investment from all sources in the development of groundwater has also grown from less than Rs.600 million (US $ 12 million) during the First Plan (1951-1956) to over Rs. 12 billion (US $ 2.4 billion) during the Eighth Plan period (1992-1997). The institutional finance for groundwater development is about 60% of the total outlay for groundwater development while public outlay is 20% and the balance is met from the private sources (Kapoor, 2000).

About 75% landholders are small and marginal farmers having less than 2 hectares of land per family. The area owned by them is about 30% of the total cultivated area in the country. At per the present norms, Government provides subsidies @ 33% to marginal and 25% to small farmers for the construction of wells and purchase of pump sets.

The Central Ground Water Board (CGWB) has established a country wide observation well network for monitoring water levels. A total decline of 4 m i.e. 20 cm/year has been noted during 1981-2000 in several states of the country. In the NCT (National Capital Territory) of Delhi, in some areas, a decline of as much as 5 to 10 m is noted. In several cases the water levels have fallen below the suction lift of existing pumping equipment.

The CGWB has categorized areas based on the present stage of groundwater development and past water level trends. Based on these criteria, about 300 blocks, all over the country, have been classified as 'over-exploited' where groundwater development is more than 100 percent and 160 blocks are 'dark' where the stage of groundwater development is between 85 and 100 percent.

Such a categorization provides guidelines to financial institutions for providing loans to farmers for the construction of additional wells. 'Grey' and 'white' areas are those where groundwater development has been 65-85% and < 65 % respectively. Overexploited areas are not worthy of additional groundwater development while in dark areas development should be undertaken with care. In grey and white areas there is a scope of further groundwater development.

According to a study carried out by the International Water Management Institute(IWMI), India is among the 50 most water poor countries. This study is based on the Water Poverty Index (WPI). India is ranked 47th in WPI while Canada, Norway, Finland, Iceland, Guyana and Surinam are among the water rich countries. At the lower end of the scale lie Ethiopia, Rwanda and Malawi. Almost all the water rich nations are in the northern hemisphere and all those with less water are in Africa. The WPI is based on resource, access, use, capacity and environment. The WPI is not an indication of only water availability but it takes into account the effectiveness of its use also.

GROUND WATER RIGHTS

Under the Indian Constitution, the land owner has an absolute right on the groundwater under his land. Therefore, a land owner can abstract any amount of water irrespective of its effects in the adjoining areas. It also means that 30% of the landless rural population do not have the water rights. Such an approach violates the fundamental rights to life of a person in a socialistic society. Therefore, the above water rights have to be modified to ensure equitable availability of water to all the citizens irrespective of land holdings.

There are also inter-state problems in the sharing of not only surface water but also of groundwater as major aquifers run across the state administrative units. Therefore, water should be recognized as a national-rather than a regional-resource which needs cooperation amongst states on these issues.

GROUNDWATER LEGISLATION

Recognizing the importance of proper groundwater development, a model bill was framed by the Government of India in 1970 and was circulated to the states. On the basis of experiences in the last two decades, the Bill was revised first in 1972 and again in the year 1996.

As a subsequence of this, the Govt. of India constituted the Central Ground Water Authority (CGWA) for the purpose of regulation and control of groundwater management and development including water quality. The CGWA has identified certain areas where for further groundwater development prior permission of the Authority is necessary.

However, it is being realized that implementation of such restrictions is difficult. The bureaucratic setup of the Government is often unable to solve the local problems without involving local people. This also requires mass awareness about the adverse effects of overexploitation and degradation pf groundwater quality. Local elected bodies, viz. village councils (panchayats) should be involved in decision making. Non-Government Organizations (NGOs) can also play an important role as people will listen to them with greater care due to their involvement in other local developmental activities, such as health and education.

GROUNDWATER QUALITY

In addition to salinity (both inland and coastal), high concentration of fluoride, arsenic, iron & nitrate are the main contaminants in groundwater. Unscientific disposal of industrial waste and sewage also causes widespread pollution of groundwater mainly in the urban and industrialized areas.

High concentration of fluoride (more than the permissible limit of 1.5 mg/l) is reported both from the hard rock (granites & gneisses) as well as alluvial aquifers. People living in such areas were drinking high fluoride waters without realizing its presence which caused various bone diseases. The cause of high fluoride in groundwater is geological, being a result of the dissolution of fluoride bearing minerals. Defluoridation plants have been installed in areas where such problems exist.

A very serious problem detected recently is the presence of As in groundwater of several districts in the Ganga river delta, covering parts of Bangladesh and eastern part of India (West Bengal). About half a million people in eight districts of West Bengal suffer from various arsenic related skin and other diseases.

Shallow aquifers in the depth range of 15 to 75 m have higher concentration of As (0.3 to 1.1 mg /l) while the deeper aquifers (>150 m_) contain negligible amount of As. The origin of As in groundwater of West Bengal (India) and Bangladesh is controversial.

However, it is regarded to be geological derived from the As rich pyrite present in peat and clay formations in the deltaic alluvium (Mukherje et al., 2002). Mitigation is difficult given the scale of the problem as well as social and economic factors. Options include use of deep groundwater (>150 m), rain water harvesting and treatment of contaminated groundwater (BGS, 2000).

Inland salinity is predominant in canal irrigated areas due to water logging. People living in these areas are also exposed to malaria and other diseases.

Over-exploitation of groundwater in the coastal aquifers of Saurashtra in Gujarat state, Tamilnadu in south India and coastal parts of Orissa in the east has led to seawater intrusion and thereby water quality deterioration.

FUTURE STRATEGIES OF GROUNDWATER DEVELOPMENT

In order to meet the future demands of groundwater, some aspects of groundwater exploration, development and management, which are necessary, are listed below:

(1) Groundwater development in water logged areas

It is estimated that about 25000 km^2 area under canal irrigation suffers from water logging and soil salinity due to high water table and lack of proper drainage. Such areas can be reclaimed by the conjunctive use of groundwater and surface water.

(2) Evaluation of Deep Aquifer Systems

Studies by the Oil and Natural Gas Corporation (formerly Oil and Natural Gas Commission) and Central Ground Water Board indicate the presence of deep aquifers (500 to 1000m) in the Ganga Brahamputra alluvium (Jones, 1987). In hard rocks areas also, there is possibility of finding groundwater at depths more than 100 m in fractured rocks (Singhal,1999). These areas are to be further explored mainly from the point of view of recharge mechanism and water quality.

(3) Groundwater Development in the Eastern parts of the Country

In the eastern parts of the country the present stage of groundwater development is estimated to be about 25% mainly due to the fragmented nature of land holding, non-availability of public funding and power supply (Kapoor, 2000). These areas are worthy of groundwater development.

(4) Inter-Basin water Transfer

The Government of India has taken up an ambitious programme of long distance transfer of surplus surface water to water-deficient areas. Necessary investigations are required to study the possible impact of inter-basin water transfer on groundwater régime and its environmental impact. This also requires consensus amongst various states and necessary political leadership.

(5) Groundwater Development in Drought Prone Areas

Drought is a recurring feature causing great miseries including stress on water availability for various uses. In view of this, high priority is required to provide safe drinking water to rural population and schools. With this in view, the Government of India, under the Accelerated Rural Water Supply Programme (ARWSP), has allocated Rs 2070 million (US $ 420 million) for the year 2003 to drill wells which will be used only during drought period as a crisis-management measure.

(6) Groundwater-Seawater Interaction

India has a long coast line stretching to about 3000 km. In certain areas, over-exploitation of groundwater has resulted in seawater intrusion in coastal aquifers. Study of groundwater-seawater interaction including quantitative assessment of fresh groundwater discharge into the sea is necessary for the optimum development of groundwater resources of coastal aquifers. The International Association of Scientific Hydrology (IASH) has also proposed recently various types of research in this field (IASH,2002).

(7) Artificial Recharge

The importance of artificial recharge including rain water harvesting is being increasingly realized to conserve the water resources. In some mega-cities, roof-top rain water harvesting has been made mandatory under the new building codes. Underground storage by artificial recharge will also prevent floods.

(8) Availability of funds

One of the problems in groundwater development has been the availability of funds especially to small and marginal farmers for the well construction.

(9) Water Quality Monitoring

The effect of natural processes and man's activities on groundwater quality needs regular monitoring for taking up timely corrective measures. Necessary mass awareness and education is also an important component

(11) Active Participation of Women

Women's participation in water resources projects especially in those involving water supply and sanitation at village level will greatly improve the implementation and sustainability of the projects.

CONCLUSIONS

The growing population of the country needs more water for various uses. However water can be constraint, if not prudently developed and managed. The real life problems facing India in the water resources sector are about the water rights including sharing of surface and groundwater resources between the various states and individuals, general shortage of water in rural and urban areas, deterioration in water quality and degradation of the environment. These aspects need due attention by administrators, scientists, engineers and social workers to ensure equitable distribution of water for the sustenance of mankind without endangering the needs of future generation.

5

Instream Flows

States that straddle the 100th meridian have long wrestled with the complex issues surrounding Great Plains and Mountain states water rights. The home state of Prairie Fire, Nebraska, has often postponed the resolution of many aspects of various struggles over water rights.

Even in the recently adjourned session of the Unicameral, days were spent attempting to balance the interests of surface water and groundwater users, given that many of Nebraska's rival basins are now fully appropriated or even overappropriated. The Nebraska Legislature did not take the first step to recognize the obligation to protect streams as a fish and wildlife resource until 1984. Even so, in the intervening years, only slightly over 2% of Nebraska's streams and rivers have received instream flow protection.

Instream flows-defined simply as the water flowing in a stream channel and its variations in flow levels across different seasons and wet, dry and normal years-have received renewed attention as the water struggles continue between surface water users and groundwater users. The subject has an enormously complex history.

Accordingly, we decided to publish a comprehensive, albeit lengthy, essay on instream flows in Nebraska. We hope that our efforts, combined with Gene Zuerlein's, will succeed in equipping our readers with the knowledge needed to effectively engage in the many water discussions that lie ahead in this, the Century of Water.-Publisher

Nebraska's obligation to protect streams and fish and wildlife resources (recognized as the Public Trust Doctrine) originated in the 6th century A.D. in Roman history. Justinian, emperor of Rome at the time, codified a law stating "By the law of nature, these things are common to mankind-the air, running water, the sea, and consequently, the shore of the sea." England adopted this Public Trust Doctrine as common law, which was subsequently exported to the original 13 colonies here in the United States.

Following the fight for independence, the Public Trust Doctrine was adopted as part of the basic laws in each state. Under U.S. Constitutional provisions, rights and obligations not specifically reserved for federal jurisdictions were transferred equally to all new states joining the union. When Nebraska accepted statehood in 1867, it accepted certain rights, common laws and obligations provided to the 13 original states and each additional state. Among others, each state accepted the common law known as "The Public Trust Doctrine (PTD).

The Constitution of the State of Nebraska states "The use of the water of every natural stream within the State of Nebraska is hereby dedicated to the people of the state for beneficial purposes" (Article XV, Sec. 5). "The right to divert unappropriated water of every natural stream for beneficial use shall never be denied except when such denial is demanded by the public interest" (Article XV, Sec. 6).

Pre-1984

The public trust doctrine revolves around the concept that government owes its citizens special duties of stewardship regarding common property resources which the state holds in trust for the public. The public trust doctrine holds that government must act as a fiduciary (one to whom property or power is entrusted for the benefit of another) in its management of the trust resources. The beneficiaries of this trust are the citizens of the state, including future generations. Common property resources, including water, fish, wildlife, air and land, are held by the states in their sovereign capacity and are subject to the public trust.

When covered wagons rolled across Nebraska in the mid- to late 1800s, there were no laws governing use of surface waters. Naturally, controversy occurred over who could use it, and 22 years after Nebraska became a state, the legislature enacted a law on March 27, 1889, which recognized and protected rights to appropriate surface waters of the state. This act established the fundamental principals to the appropriative system: first in time, first in right. Six years later, the Comprehensive State Code Act of April 4, 1895, created the State Board of Irrigation and firmly established the current appropriation system for surface water rights. Eventually, this state agency became known as the Department of Water Resources, and in more recent times, the Nebraska Department of Natural Resources (NDNR).

While the appropriation system worked well during the development years of the state, it may not be the best for future generations because much more than three or four instream flow purposes need to be protected. It is pretty much recognized today that human welfare is dependent upon ecological health and that when rivers no longer support living things, they will no longer support human affairs as well. How much can we depart from the natural flow patterns of any particular river and still retain enough semblance of the natural character to meet ecosystem needs? Protecting rivers should really be about preserving the processes and functions needed to sustain ecosystems and the many amenities they provide to society as a whole.

Although pumping groundwater was in its infant stages in the 1920s, the "correlative use doctrine" was legally recognized in Nebraska in 1933. Basically, correlative use means that in times of water shortages adjacent users must share and share alike. Management of groundwater resources up to this point has, for the most part, been relegated to 23 political subdivisions of the state called Natural Resources Districts (NRDs) with some oversight by the Nebraska Department of Natural Resources. These NRDs, organized on a watershed basis in 1974, are charged with regulating groundwater pumping. Up until passage of LB 962 in 2004, two sets of laws, one governing surface water appropriations and one governing groundwater pumping, existed, with neither recognizing the conjunctive aspect of both sources of water.

Both surface and groundwater resources have been extensively developed since 1889. In water year 1975, 93.2% of all water used in the state was for irrigating 5,400,000 acres. Groundwater irrigation accounted for 66.7% and surface water irrigation accounted for 26.5%. By 1986 approximately 7,200,000 acres were under irrigation. Nine years later (1995), more than 7,448,700 acres were irrigated. Groundwater provided 82.6% and surface water provided 17.4% of the water consumptively used. During 2001, there were 8,175,000 acres irrigated in Nebraska.

While irrigation was expanding in Nebraska, other authorities recognized the potential risk and were advocating change. The 1976 "Missouri River Basin Commission's Platte Level B Study" boldly stated:

"Progressively greater use of groundwater for irrigation and other uses as well as direct uses of the surface waters, particularly for irrigation, are having adverse effects on wetlands, natural lake levels and on stream flow regimes. These adverse effects will continue, unless there is recognition in state law and policy of the need to maintain certain minimum stream flows to ensure the maintenance of the fishery resources, wildlife, and other environmental resources."

In 1976, portions of 21 streams in northeastern Nebraska became totally dry and fish kills during the summer were the most numerous in recorded history. Drought, excessive stream withdrawals and groundwater mining contributed to the lowering of stream levels. For example, on the dry Platte, some surface water appropriators reverted to digging holes in the bed of the river so they could get their stingers to suck up the last drops of subsurface water for irrigation. In advocating for an instream flow law, one concerned citizen stated, "The Nebraska environment must not be blindly sacrificed in the name of economic development." Another citizen stated, "The past few years have demonstrated that water, like any other natural resource, is exhaustible. It is evident from dewatering of Nebraska streams that the water laws of Nebraska have not kept up with the needs of the environment."

In 1978, the Legislature passed LB 957 and LR 300, directing the Natural Resources Commission (now Nebraska Department Natural Resources or NDNR) to accelerate state water planning. This led to nine Policy Issue and one Summary and Review Policy Issue Studies. One of these studies, "Instream Flows," specifically identified streams and rivers which contained unappropriated water. It also recommended flows required for: fishery needs in 13 streams/rivers, canoeing in six rivers, compact flows for the Little Blue and Big Blue, on-stream hydroelectric generation flows for four rivers and Minnechaduza Creek, recharge of Lincoln's well field along the lower Platte, and flows for whooping crane habitat on the Niobrara River. In the summary section of this report, it distinctly states that "Nebraska's present system for acquiring and administering appropriative rights was adopted in 1895 and has undergone little change ... under Nebraska law an appropriator may impound or divert the entire flow of a stream if necessary to satisfy his water right regardless of the impact on stream flow values" (emphasis added).

Post-1984

In 1984, after almost a decade of debate and study on the need for instream flows and 95 years after Nebraska water laws were first adopted, the Legislature passed LB 1106. For the first time in state history, a statutory mechanism to protect instream flows in streams and rivers with unappropriated flows materialized. This legislation was significant because it finally recognized that surface water appropriations for instream flow values of fish, wildlife and recreation could be obtained. The Legislature gave responsibility (but no money) for obtaining instream flows to the Nebraska Game and Parks Commission and the 23 Natural Resource Districts in the state. During the 23 years since 1984, only 255 miles (2.1%) of Nebraska's 12,371 miles of streams and rivers have been protected with instream flow appropriations.

Protected reaches include eight miles on Long Pine Creek (a tributary to the Niobrara River), 147 miles on the middle

Platte and 100 miles on the lower Platte River. One has to wonder where the state of Nebraska would be now in protecting its public trust waters if each NRD had just filed for one instream flow appropriation to protect the main river in their district during the past 23 years. While all initial applications for instream flow protection have been contentious, many river basins are now fully or overappropriated. Most basins have no instream flow appropriations for fish, wildlife or recreational resources and consequently many of the amenities flowing water could provide to our society are in jeopardy unless actions are taken to change state policy to better protect stream resources for future generations.

In 2004, the Legislature passed LB 962 to make the NDNR and 23 NRDs more proactive in anticipating and preventing conflicts between groundwater and surface water users. In fully or overappropriated basins, the NDNR must work with the NRDs to develop and implement integrated management plans. The NDNR must also annually evaluate each remaining river basin to determine if it is fully appropriated. However, unless there is an instream appropriation in the basin, there is no mandatory requirement under LB 962 for NDNR's evaluations to protect public trust instream resource values. All they can do is listen to concerns about the many other amenities a river provides.

If streams continue to be depleted, biological opinions regarding instream flows needs for threatened or endangered species may very well be another factor to consider in fully appropriated river basins.

The fact remains that until adequate flows through an instream appropriation are actually granted, new surface water appropriations and groundwater pumping permits are likely to continue to be issued until a stream is declared "fully appropriated." Since the majority of streams and rivers do not have instream flow protection, once a fully appropriated status is declared for a basin, water will have to be leased (or someday purchased) similar to what is being contemplated for the fully appropriated Republican River Basin and major portions of the Platte River Basin.

A new provision of LB 962 allows leasing of the consumptive portion of a surface water appropriation (e.g., an irrigation diversion appropriation) for instream flows or other uses within a basin for a period of time up to 30 years. This is a start, but it may also be more appropriate for those who are going to have to provide tax dollars to allow for perpetual purchasing of such water in the long term.

During 2006, when assessing hydrological connections along streams and rivers, the NDNR adopted a hydrologically connected area regulating rule called "10-50." Under this rule, groundwater supplies are considered to be hydrologically connected to a stream or stream reach if the groundwater is draining to the stream.

In accordance with NDNR Rule 457 NAC 24.001.02, the area within which groundwater is hydrologically connected to a stream is the area within which "pumping of a well for 50 years will deplete the river, or a base flow tributary thereof, by at least 10% of the amount pumped in that time." Wells within the geographic boundary defined by the 10-50 rule can legally be regulated to prevent a basin from becoming overappropriated, but wells outside the 10-50 line may not be regulated even though they may also contribute to depletions of stream flows.

The provision to allow leasing of the consumptive part of a surface water appropriation will have merit in the recently approved 2006 Platte River Recovery Implementation Program, where the state of Nebraska will have to pick up the costs of replacing any depletions from designated target flows which occurred after July 1, 1997, because of the Cooperative Agreement signed by the governors of Nebraska, Wyoming, Colorado, and the U.S. Department of the Interior Director on this date. While the purpose of the Cooperative Agreement is to improve and maintain habitat for the endangered whooping crane, interior least tern and piping plover (threatened) in the critical reach of the central Platte River, and endangered pallid sturgeon in the lower Platte River reach, better flows in the central and lower Platte will help many other species and their habitats as well, not to mention other amenities to communities and citizens living along these reaches. Tough decisions remain to be made to fulfill this obligation.

It seems prudent that the state of Nebraska could save taxpayers a lot of money if rivers and streams were protected before they become fully or overappropriated. Since the waters of the state are public trust resources belonging to all citizens, it begs the question: How well have the state of Nebraska and its officials protected the public trust?

The United States has passed through four eras of public land and resources management, each with its own legislative policies and environmental ethic.

There was settlement and development of the original public domain (1789-1834); next, a public land resource stewardship with a conservation ethic began with President Theodore Roosevelt in the early 1900s; after World War II there was ensuing national growth, including many water development projects such as dams (1941-1962); last, there was the environmental law era of 1962-1990. These four eras paved the way for the present era of watershed restoration and collaborative stewardship.

State-of-the-art understanding of the dynamic nature and ecology of streams and rivers and their importance to providing ecosystem goods and services to mankind has increased tremendously in the last two decades. While the task of protecting instream flows to sustain these uses is a daunting one, those responsible for protecting the public trust must join hands with fellow citizens if future generations are to have any hope of enjoying what present generations enjoy as part of their quality of life here in Nebraska.

The need for ecosystem management of large rivers and their floodplains has never been greater. Significant economic advantages, as well as increased biodiversity, can result from restoration of impaired systems.

Because ecosystem goods and services rendered by rivers and streams were not well understood historically, they were underappreciated. As a society, we are just beginning to awaken to the true cost of ignoring them. In many cases, it will be cheaper to restore them than to exploit them beyond sustainability.

Nebraska's surface and ground waters have been extensively developed for many purposes, but irrigated agriculture is the largest consumptive user of water in the state. Some would construe the development and over-exploitation of our surface and groundwater to be a "Tragedy of the Commons."

In this context, state property right laws (e.g., when a surface water appropriation to divert is issued and perfected, it becomes a property right, or the unlimited issuance of irrigation well permits to pump groundwater by NRDs) have failed to safeguard important public trust natural resources which belong to all citizens.

Together, they are impacting many stream and riverine natural resources which contribute to the physical, social, biological and economic aspects of Nebraska's economy. Collectively, some people refer to these resources as ecosystem goods and services. Essentially, they are environmental flow regimes of a particular magnitude, frequency and timing necessary to ensure that a river system remains environmentally, economically and socially healthy for its citizens. Environmental flow regimes are heartbeats for rivers and should not be flatlined.

Although this sounds bleak for managing public trust natural resources on a sustainable basis, much work remains to be done. There are some rivers with unappropriated water left which can be protected with instream flows. On May 24, 2006, the Nebraska Game and Parks Commission (NGPC) Board of Commissioners passed a resolution directing staff to protect existing instream flows already granted and to secure instream flows on streams with high-value natural resources, particularly the heartbeat of the Niobrara River with its high-quality tributaries.

Should more and more basins become fully or overappropriated and streams and rivers dry up more frequently, the state of Nebraska could very well have to lease or purchase water back to restore and protect public trust/ public interest values. Depending upon assumptions made and analyses conducted, it could cost millions of tax dollars to accomplish this. Since surface water is a public resource regulated by the NDNR and groundwater is a public resource regulated locally by NRDs, it remains to be seen whether or not the fortitude to prevent or solve overexploitation of both exists.

Funding for NRDs primarily comes from local property tax levees, but state officials may not be open to raising state taxes on all citizens in the future to bail out local NRDs whose decisions result in overexploitation of groundwater in their basins, thus resulting in depletion of hydrologically connected stream/river flows and, subsequently, the loss of current and future ecosystem goods and services.

More than three out of four Americans participate in active outdoor recreation each year. Yet many people fail to realize that having fun and staying healthy in the outdoors is essential to the growth of our economy. According to the Outdoor Industry Foundation, the active outdoor recreation economy contributes $730 billion annually to the US economy.

During 2006, an independent economist from ECONorthwest by the name of Ernie Niemi out of Eugene, Ore., evaluated the potential to grow Nebraska's economy by capitalizing on natural-resource-related, amenity-driven growth in Nebraska. His report, entitled "Natural-Resource Amenities and Nebraska's Economy: Current Connections, Challenges, and Possibilities," is available from the Nebraska Game and Parks Commission. His conclusion: Natural-resource-related amenities may be able to stimulate economic growth through four mechanisms.

* Improving the quality of life by making the state attractive to entice entrepreneurs.
* Improving feedback to the farm sector by increasing off-farm job opportunities.
* Expanding recreational and other commercial uses of natural resources.
* Protecting environmental values by sustaining species and keeping the web of life together.

A significant proportion of Nebraska's future economic vitality could potentially revolve around its streams and rivers if natural attributes are protected enough so they can generate amenities citizens are seeking. Examples where this is happening include ecotourism on the central Platte for bird-watching; ecotourism on the Niobrara for canoeing and tubing; and ecotourism on the Missouri River from boating and many other recreational uses.

The Missouri River has an active recovery program for the endangered pallid sturgeon, least tern and piping plover, as well as for the diversity of habitats needed to support a healthy riverine ecosystem. A healthy Missouri River means that historically forgotten amenities will be more prominent in the future to help grow Nebraska's (and other basin states') economy by paying dividends to many Nebraska citizens seeking fun places to connect with nature, the outdoors and rivers.

While Nebraska's instream flow provisions are the most restrictive among western statutory laws, they could be improved upon considerably. Adequate instream flows benefit both ecological and economic interests.

They are the essence of what makes a river and contribute to many ecosystem goods and services, including filtration and dilution of sewage and associated effluents for water quality; livestock watering; increased land values; cooling water for electrical generating plants; hydro-generating plants; drinking water sources and groundwater recharge; recharging riparian wetlands; subirrigating wet meadow grasslands; fish, wildlife and recreational uses; and many, many other things important to the quality of life here in Nebraska.

However, the question remains: when stream and river flows can legally be diverted or diminished by some combination of surface water diversions and/or groundwater depletions to the point where the lifeblood is drained out of them and they no longer flow, has the state of Nebraska fulfilled its obligation of protecting the public property rights embodied by the public trust? Has the public interest of present and future generations been served?

6

Augmentation of Groundwater Resources

Over the last 20 years there has been an enormous increase in the use of ground water resources all over the world. Besides providing an important source of water for domestic and agricultural use, groundwater also plays a key role in supporting and maintaining the livelihoods of the poor.

This is because groundwater can be accessed relatively easily and cheaply and its quality is generally high. The intensity of groundwater exploitation has raised concerns that groundwater use and the livelihoods it supports-may not be sustainable.

Symptoms of over-exploitation include declining water levels and increasing competition between users for scarce supplies. Now a days due to over exploitation of ground water for irrigation purposes, the water table is falling by about 20-25 cm per year in good ground water quality regions of Punjab, Haryana & Gujarat (Source: CGWB). Therefore there is urgent need for recharging of groundwater resources.

Artificial Recharge

Artificial recharge is an important aspect of groundwater management as it provides storage space free of cost avoids evaporation losses and allows use of stored water during dry period. Artificial recharge is achieved by three methods, namely by:

1. Spreading

2. By induced recharge and
3. By injection

The suitability of a particular method is based on the hydro geological conditions, quality of source water and proposed use of recharged water. These methods are explained below:

Spreading Methods

This method is similar to that of surface irrigation with the difference that in recharge operation the water is allowed to stand on the field for a much longer period.

- Getting the water into the ground.
- Getting the water down through the vadose zone to the aquifer.
- Getting the water to move through the aquifer away from the infiltration site without undue build-up of groundwater mounds.

Thus, for successful infiltration of recharge systems, the following conditions are desirable:

- Surface soils must be sufficiently permeable to maintain high infiltration rates.
- Vadose zones must be permeable and free from clay layers or other finer materials that could restrict downward flow of water and produce perched groundwater that could back up into infiltration basins.
- Aquifers must be unconfined and permeable and thick enough to avoid excessive rise of groundwater mounds.
- Groundwater tables must be deep enough to allow some rise in the water table, or the water table must be controlled with wells or drains that take the recharge water out of the aquifer again.

Keeping in view the above-mentioned conditions and hydro geological set-up of Punjab State, the following areas offer suitable conditions:

- Kandi belt adjoining the Shivalik system where alluvial plains exist.

- Area of old river courses, which can be identified with the help of Landsat data.
- Areas around flood plains (in and along stream beds).
- Areas different than the above three categories and where unconfined aquifers predominate.

In all the four categories, sometimes, sandy soils at top may be covered by clay or other fine textural material. This situation occurs in flood plains or other alluvial deposits. In case the layer of fine top materials is not too thick, it can be removed and the infiltration basins can be excavated into the coarse underlying deposits as shown in figure below. Surface soils also play an important role in controlling the infiltration rates.

According to size soils are classified as clay (<2 microns), silt (2-50 microns), and sand (0.05-2 millimeters). Particles larger than 2 mm in diameter are classified as gravel (fine, coarse, cobbles, and boulders). Usually infiltration basin soils should be sandy loams or coarser to produce acceptable infiltration rates (for example, more than 20 cm/day). Coarse sands and gravels can give very high infiltration rates.

Table: Showing Soil Texture and Estimated Rate of Infiltration

(Modified after Todd, 1985)

Infiltration Rate	Soil Texture
High (Greater than 5cm/hr)	Coarse sand, sand, fine sand, loamy sand, loamy fine sand, coarse sandy loam.
Intermediate (Between 5 cm/hr and 1.5 cm/hr)	Sandy loam, fine sandy loam.
Low (Less than 1.5 cm/hr)	Silt loam, sandy clay loam, clay loam, Silty clay, sandy clay, clay.

Methods of water spreading involve:

- Construction of Basins
- Stream channels
- Ditch and furrows

- Flooding
- Irrigation
- Pits and shafts

These methods have been discussed below in relation to their utility for Punjab State .

Basin Method: In this method, basins are constructed by excavation or by building dikes or levees. Sizes and shapes are adjusted to fit topography, land ownership and available land. Depending on topography, basins can have a surface area of 0.1 hac (0.25 acre) or less to 10 hac (25 acres) or more. Often, the first few basins are "sacrificed" as presedimentation facilities.

Water is then released to basins for infiltration. Water released should have minimum sediment load so to reduce deposition and scaling cf basin surfaces. This can be accomplished by three ways:

- Diverting water to basins during non-flood periods when suspended sediment is low.
- Adding chemical flocculating agents to water to remove sediment.
- Providing sedimentation basins to hold water before releasing it into recharge basins.

Basin designs vary widely depending upon local conditions. Basin width should be sufficiently wide to accommodate scrappers to remove sediment periodically. Side slopes should be relatively steep so as to minimize sediment deposition.

In actual infiltration basins, suspended material in the water (silt, clay, organic particles, etc.) accumulates on the bottom. Also bacterial and algae growth in the basins produces algae crests and organic deposits on the bottom. These accumulations reduce infiltration rates, typically if the water contains high suspended solids content and the basins are flooded for a long time. Thus, the hydraulic conductivity 'K' of the bottom soil gives an idea of maximum sustained infiltration rates in the basin. Order of magnitude of hydraulic conductivity K for different soils are (in m/day):

Clay soils (surface)	0.01-0.2
Deep clay deposits	10-8-10-2
Loam soils (surface)	0.1-1
Fine sand	1-5
Medium sand	5-20
Coarse sand	20-100
Gravel	10-1000
Sand and gravel mixes	5-100
Clay, sand and gravel mixes	0.0001-0.1

These values give a general idea of K-ranges for different soils, hence, of the maximum infiltration that can be obtained to get more reliable estimates.

In Punjab infiltration basins can be constructed close to rivers, and in areas of old river courses, and in wide channels. Studies by Central Ground Water Board have shown that basins of 250 m X 60 m size and having depth of 2 m to 5 m can be constructed to carry artificial recharge.

Stream Channel Method: Water spreading in stream channels involves operations to increase the time and area of water contact. Activities typically require upstream storage facilities to regulate stream flows and channel modifications to enhance infiltration. Ideally, upstream reservoirs should limit flows to rates that do not exceed the absorptive capacity to downstream channels. Types of stream channel improvements include:

- Widening leveling, scarifying or ditching of the channel.
- Permanent low check dams, which do not cause flood hazard.
- Temporary low check dams consisting of stream bed materials.
- L shaped finger levees constructed by bulldozer at the end of high-stream flow season.

In Punjab, this method is particularly suitable for Kandi area of District Ropar, Hoshiarpur, and Gurdaspur.

Ditch and Furrow Method

In this method continuous ditches and furrows are constructed throughout the field, which facilitates the infiltration rate by providing more surface area. Ditches and furrows are adaptable to irregular terrain where other methods of water spreading would not be feasible. This method is not suitable for Punjab State and is mainly practicized in hilly regions with uneven topography.

Flooding Method: In flat land areas, water can be simply released over large area and allowed to infiltrate. Velocities of flow should be minimal to avoid soil movement. Usual practice is to leave vegetation of that area undisturbed. This method is the least costly of all spreading methods, provided the land is available without cost. For maximum efficiency, hard labour should be employed to distribute the water over area as wide as possible. This method has very limited scope in a state like Punjab where most of land is cultivated.

Irrigation Method

Water can be recharged on irrigated lands by three methods:

" Applying excess water during non-irrigation seasons.

- Applying excess water during irrigation (However, this can have adverse effects on leaching salts and removing soil nutrients).
- Keeping irrigation canals full.

This method is also not suitable to the conditions of Punjab except in last option i.e. keeping the irrigation canals full in central parts of Punjab having sweet water (District Ludhiana, Jalandhar, Kapurthala, Amritsar, Gurdaspur). However, canal network is very little in Bist Doab Tract (Ludhiana, Hoshiarpur, Jalandhar). As such this method has also very little scope in Punjab.

Pit Method

Pits are excavated into permeable formations, which serve as ideal facilities for artificial recharge. Abandoned gravel pits are most economical for this purpose. This method has also very little application in Punjab. Induced Recharge:Induced recharge occurs where groundwater is pumped from a location near a surface water body so that lowering of groundwater level induces water to enter the ground from the surface source.

The method is effective in permeable formations, which hydraulically connected between the stream and the aquifer. The amount of water depends on: pumping rate, permeability, type of well, distance from stream and natural groundwater movement. The stream velocity should be sufficient to prevent sediment deposition from sealing the streambed.

Injection Methods: Such methods are of particular interest where over drafting occurs in confined aquifers and areas are located close to major canals. As water is available during monsoons, injection methods are suitable on account of clean waters available close to sites. Such methods can be tried areas having confined aquifers and suffering from over drafting of groundwater.

Other Measures:

- The other measures aim at reducing the ground water overdraft. It is suggested that in severely over draft areas, new cropping patterns may be introduced. Crops, which require less water than the existing crops, shall have to be promoted. e.g. wheat Rice-Rotation practice should be replaced in Punjab and Haryana as early as possible.
- In critical areas, legislative measures can be introduced. Pumping patterns can be regulated keeping in view the available water supply in the canals.
- There is need to educate the farmers regarding optimum water utilization aspects, irrigation planning etc. Public awareness regarding water conservation measures should be created through Celebration of 'National Water Day' and with the help of awareness programmes through people's participation and voluntary organizations.
- Recycling of water and reuse of water should be promoted especially in those areas where industries consume groundwater.

By utilizing one or a combination of above-mentioned methods depending on its suitability to a particular area we can regenerate our groundwater resources to a great extent.

CONSTRUCTED WETLAND TECHNOLOGY

The population of Punjab has increased rapidly over the past fifty years. This has put a great pressure on natural resources. This, combined with low quality of sanitation and lack of awareness among the people of rural Punjab, has resulted in degradation of the natural environment. Though water supply has increased in rural areas but wastewater collection and disposal facilities have not developed concurrently.As a result most of the drains and storm water flows into the natural village ponds. Flow of cow-dung & soil into the ponds along with storm-water adds to the pollution load resulting in degradation of the water quality. Hence these ponds are no longer fit for their traditional uses, like, bathing of humans & cattle and there is a likelihood of water-borne diseases being spread.

The conventional wastewater treatment systems (like Activated Sludge Plants and Trickling filters) do not present a viable option in the rural areas of Punjab owing to the high cost of construction and maintenance and the lack of technical people to operate them. This situation demands the need to find economical and sustainable solutions for water quality improvement in the ponds situated in the rural area of Punjab. A project has been taken up by Punjab State Council for Science &Technology in collaboration with M/s CH_2M HILL Canada to promote an alternative low cost technology for sustainable water treatment. The project has been taken up in villages Peepal Majra (District Ropar) and Shekhupur (Distt. Patiala).The project sites face similar problems of degrading water quality due to influx of sullage and storm-water in the village ponds.

The Technology

Constructed surface flow treatment wetlands are typically shallow impoundments planted with emergent, rooted vegetation. Water flows overland through the wetland and primarily above the sediment surface. They contain monoculture or diverse plant communities that are more adaptable under changing season and water quality conditions. These wetlands may be planted manually or may be naturally colonized by volunteer plant communities. The plants in constructed wetlands are not harvested to remove nutrients, rather the microbial flora that is attached to the plants have the natural capacity to remove biodegradable organics and nitrogen efficiently and reliably. Metals and phosphorus can be restored in plant materials and wetland sediments. The treatment wetland can be designed to regulate water depth and residence time. Much of the treatment that occurs in wetlands is a result of sediment trapping and microbial action rather than plant uptake. Therefore, these systems continue to function during winter and are not affected by temperature due to buffering effects of the sediments.

The following benefits can be derived using treatment wetlands:

- Less expensive to construct than traditional wastewater treatment systems.
- Require less maintenance and less expensive to operate than traditional systems.
- The discharge from treatment wetlands can increase storage in ground water aquifers through combination with rapid infiltration basins.
- With proper design, portion of treatment wetland may provide additional wildlife habitat, as well as recreational opportunities such as bird watching, picnicking and fishing.
- The system can be combined to provide additional economic benefits such as preparation of handicrafts from plant biomass.

Existing Conditions

The village Pipal Majra currently has two ponds, which receive wastewater from the village. There are 192 households in the village with a population of 1,200 discharging their sullage into these two ponds. One pond with an area of 0.58 ha has been identified for implementing the project. About 50% of people discharge their sullage in this pond, with rest of the community discharging their effluent to the other pond. Most of the houses have open drainage system Wastewater from overflow of soak-pit latrines and domestic waste flows to the pond through these drains.

Cow-dung is dumped near the banks and finds its way into the pond along with the sullage and the storm water through the open drains.The local community uses the pond for cattle bathing, which also contributes to the pollution load. There is a cremation ground on the west bank and an existing Samadhi contributing to the pollution. The change in the land use pattern of the village has resulted in a decrease in the area of pond (20-25%), resulting in its reduced carrying capacity and degraded water quality.

The Proposed Project

The proposed project is designed for an expected population of 1,500 and a flow of 0.15 MLD in the year 2011. The System comprises of Primary Sedimentation Tanks followed by Treatment Wetlands.The sullage flow is routed through the open drains into the treatment pond through four primary sedimentation tanks, each with detention period of 4 hours, followed by Bar-Screens.The suspended solids reduction in the sedimentation tanks is expected to be between 20%-70% with a consequent decrease in the BOD of the sullage. The water from the Treatment Wetlands will flow in to the Treated Water Pond.

For the current project the existing pond has been divided into two zones: the 'treatment zone' and the 'treated water pond' separated by a berm. The slope stabilization on the berm is being done by a layer of compacted earth and a layer of top-soil over it with vegetation for erosion protection. The area of the treatment zone is 2,100 m2 or 0.21 ha with a width of 14m and length of approx. 150 m giving a length to width ratio of approximately 10:1 with continuous feeding of effluent into the treatment pond. The proposed treatment zone consists of deep zones for even distribution of flow along the width of the treatment pond. The depth of the water is 0.30 m with cattails (Typha) as the predominant plant species in the treatment zone. An emergency overflow exit is provided at a level of 0.6 m above the free water level for excess flows in monsoon season. The water is treated in the treatment zone by both physico-chemical and biological processes with vegetation providing surfaces for microbial activity. The effluent from the treatment zone is further aerated by routing it through a cascade before entering into the treated water pond.

Expected Outcome

The treated water will be finally discharged into the treated water pond, which would be used by the villagers for cattle bathing and washing. Steps and ramp are been provided to allow the villagers and the cattle to enter the pond. On successful completion of the project the water of the pond would be treated to a BOD of <30 mg/l and will hence, provide a better environment and cleaner water to the villagers. The treated water pond is also expected to be used for fish farming.

SPRINKLER SYSTEM OF IRRIGATION

As irrigation accounts for maximum consumption of fresh water resources, therefore proper attention is required to conserve this water. Since surface method of irrigation is most commonly practiced which contribute major portion of it to evaporation and seepage losses. This method of irrigation should be avoided and method like sprinkler irrigation method should be used to have maximum irrigation efficiency.

Sprinkler Irrigation System

In the sprinkler method of irrigation, water is sprayed into the air and allowed to fall on the ground surface somewhat resembling rainfall. The spray is developed by the flow of water under pressure through small orifices or nozzles.

The pressure is usually obtained by pumping with careful selection of nozzle sizes, operating pressures and sprinkler spacings, the amount of irrigation water required to refill the crop root zone can be applied nearly uniformly at a rate to suit the infiltration rate of soil, thereby obtaining efficient irrigation.

Due to higher installation and operational cost the sprinkler system is currently considered beneficial mainly to high value cash crops.

Design of the System

The system is designed after considering the agro-climatic like soil, plant, root-depth and climate etc. These factors enable one to develop a clear understanding of:

(a) Water application rate.

(b) Depth of application.

(c) Period of application.

Suggested maximum application-rates for sprinklers for average soil, slope and tilth.

Sr. No	Soil texture and profile	0.5% slope cm/hr	5.5% slope cm/hr	8-12% slope cm/hr	12-16% slope cm/hr
1.	Coarse sandy soil to 2 m	5.0	3.7	2.5	1.3
2.	Coarse sandy soil over more compact soils	3.7	2.5	2.0	1.0
3.	Light sandy loams to 2 m	2.5	2.0	1.5	1.0
4.	Light sand loams over more compact soils	2.0	1.3	1.0	0.8
5.	Silt loams to 2 m	1.3	1.0	0.8	0.5
6.	Slit loams over more compact soils	0.8	0.6	0.4	0.3
7.	Heavy textured clays or clay loams	0.4	0.3	0.2	0.1

Source: Adopted from Hurd (1969)

Adaptability of Sprinkler system:

- Sprinkler irrigation can be used for almost all crops (except rice and jute) and on most soils. It is however, not usually suitable in very fine textured soil (heavy clay soils) where the infiltration rates are less than about 4 mm per hour.
- The flexibility of the sprinkler equipment and its efficient control of water application make this method adaptable to most topographic conditions without extensive land

preparation and leveling. This saves considerable cost and does not disturb the soil.

- The amount of water can be controlled to meet crop needs and light application can be made efficiently on seedling and young plants leading to better germination or increased yield per hectare.
- Soluble fertilizers, herbicides and fungicides can be applied in irrigation water economically with little extra equipment. The fertilizer is distributed evenly without waste to the root system for immediate and effective response by the crops.
- Saves water thus irrigating more land. Approximately 54% to 71% of water applied by surface irrigation is lost through deep seepage and evaporation. The sprinkler system supplies only the required amount of water for crop root zone. The saved water can irrigate 25% to 33% additional area.
- Helps to grow more crops per year as three or even four crops per year can be grown. Crops can be planted at the right time without waiting for seasonal rains to achieve maximum growth.
- Helps in saving of land as there is no need to contract bunds or water channels hence more land under crops.
- Reduce pests as it sweeps away all pests present on the surface of leaves.
- Easier harvesting of root crops because the soil remains friable and uncompacted.
- Low running cost as compared to that of other lift irrigation system.
- Well suited for sports grounds, parks and gardens.

Disadvantages of the Sprinkler System:

- Higher initial cost.
- High power requirement
- Not usually suitable in very fine textured soil (heavy clay soils) where the irrigation rates are less than about 4 mm per hour.

Cost concept for sprinkler irrigation system, (2001)
Sprinkler system

Cost of constructing well and installing motor = Rs. 1.00 lac.
Cost for electricity connection = Rs. 0.10 lac.

Sprinkler System Set

(a) Cost of aluminum set. = Rs. 0.50 lac

(b) Cost of PVC set = Rs. 0.30 lac

Each Set Contains:

(a) 40 Main / lateral lines (30 pipes for supply, 10 for spraying)

(b) 10 nozzle sets.

(c) 1 T-Section

(d) 2 Bands

(e) 1 End Joint

Area covered = 15 feet on both sides of the lateral lines.

Total time to cover an area of one acre = 12 hours excluding shifting time.

With each set one can cover maximum area = 10 acres

Operation and Maintenance of Sprinkler Systems

Proper design of a sprinkler system does not in itself ensure success. The System should be operated in keeping with good irrigation practice. It should be ensured that the prime mover and the pump are in alignment, particularly in the case of tractor-driven pumps. For these the drive shaft as well as the pump shaft should lie at nearly the same height to prevent too great an angle on the universal shaft. Service and installation procedures in respect of the pump and power units should be strictly observed.

While laying the main and lateral pipes, always begin laying at the pump. This necessarily gives the correct connection of all quick coupling pipes. While joining couplings, it is ensured that both the couplings and the rubber seal rings are clean.

In starting the sprinkler system, the motor or engine is started with the valves closed. The pump must attain the pressure stated on type plate of otherwise there is a fault in the suction line. After the pump reaches the regulation pressure. The delivery valve is opened slowly. Similarly, the delivery valve is closed after stopping the power unit.

The pipes and sprinkler-lines are shifted as required after stopping. Dismantling of installation takes place in the reverse order to the assembly described above.

Maintenance: A sprinkler system, like any other farm equipment, needs maintenance to keep it operating at peak efficiency. Parts of the system subject to the most wear are in rotating sprinkler heads, the pumping set, the couplers and the pipeline. General principles regarding the maintenance of the pipes and fittings and sprinkler heads are given below:

Pipes and Fittings

The pipes and fittings require virtually no maintenance but attention must be given to the following procedures:

(a) Occasionally clean any dirt or sand out of the groove in the coupler in which the rubber sealing ring fits. Any accumulation of dirt or sand will affect the performance of the rubber sealing ring.

(b) Keep all nuts and bolts tight.

(c) Do not lay pipes on new damp concrete or on piles of fertilizer. Do not lay fertilizer sacks on the pipe.

Sprinkler Heads

The sprinkler heads should be given the following attention:

(a) When moving the sprinkler lines, make sure that the sprinklers are not damaged or pushed into the soil.

(b) Do not apply oil, grease or any lubricant to the sprinklers. They are water lubricated and using oil, grease or any other lubricant may stop them from working.

(c) Sprinklers usually have a sealed bearing and at the bottom of the bearing there are washers. Usually it is the washers that wear and not the more expensive metal parts. Check the washers for wear once a season or every six months-this is especially important where water is sandy. Replace the washers if worn.

(d) After several season's operation the swing arm spring may need tightening. This is done by pulling out the spring end at the top and rebending it. This will increase the spring tension.

In general, check all equipment at the end of the season and make any repairs and adjustments and order the spare parts immediately so that the equipment is in perfect condition to start in the next season.

Storage

The following points are to be observed while storing the sprinkler equipment during the off season:

(a) Remove the sprinklers and store in a cool, dry place.

(b) Remove the rubber sealing rings from the couplers and fittings and store them in a cool, dark place.

(c) The pipes can be stored outdoors in which case they should be placed in racks with one end higher than the other. Do not store pipes along with fertilizer.

(d) Disconnect the suction and delivery pipe-work from the pump and pour in a small quantity of medium grade oil. Rotate the pump for a few minutes. Blank off the suction and delivery branches. This will prevent the pump from rusting. Grease the shaft.

Protect the electric motor from the ingress of dust, dampness and rodents.

Trouble Shooting

The following are the general guidelines to identify and remove the common troubles in the sprinkler systems:

1. Pump does not prime or develop pressure

   (a) Check that the suction lift is within the limits. If not get the pump closer to the water.

(b) Check the suction pipeline and all connections for air leaks. All connections and flanges should be airtight.

(c) Check that the strainer on the foot valve is not blocked,

(d) Check that the flap in the foot valve is free to open fully.

(e) Check the pump gland (s) for air leaks. If air leaks are suspected tighten the gland (s) gently. If necessary repack the gland (s) using a thick grease to seal the gland satisfactorily.

(f) Check that the gate valve on the delivery pipe is fully closed during priming and opens fully when the pump is running.

(g) Check that the direction of rotation of the pump is correct.

2. Sprinkler do not turn

(a) Check pressure

(b) Check that the nozzle is not blocked. Preferably unscrew the nozzle or use a small soft piece of wood to clear the blockage. Do not use a piece of wire or metal as this may damage the nozzle.

(c) Check that the sprinkler bearing is quite free and smooth. During operation, the sprinkler can usually be pushed down towards the riser pipes so that the water pressure flushes out the bearing. If the bearing is still stiff dismantle and then clean it. Do not use oil, grease or any lubricant.

(d) Check the condition of washers at the bottom of the bearing and replace them if worn or damaged.

(e) Check that the swing arm moves freely and that the spoon which moves into the water stream is not bent by comparing it with a sprinkler that is operating correctly. If it is bent then carefully bend it back into position.

(f) Adjust the swing arm spring tension. Usually it should not be necessary to pull up the spring by more than about 6 mm.

3. Leakage from coupler or fittings

The sealing rings in the couplers and fittings are usually designed to drain the water from the pipes when the pressure is turned off. This ensures that the pipes are automatically emptied and ready to be moved. When the pump is first started and before the pressure has built up in the system the seals may give a little leakage. With full pressure in the system the couplers and fittings will be effectively leak-free. If, however, there is a leakage, check the following:

(a) There is no accumulation of dirt or sand in the groove in the coupler in which the sealing ring fits. Clean out any dirt or sand and refit the sealing ring.

(b) The end of the pipe going inside the coupler is smooth, clean and not distorted.

(c) In the case of fittings such as bends, tees and reducers ensure that the fitting has been properly connected into the coupler.

In addition to water conservation the use of sprinkler irrigation reduces the runoff of soil and may reduce the runoff of pesticides and nutrients from irrigated cropland and their transport to surface waters and hence its use should be promoted as much as possible.

PITCHER IRRIGATION-A SUB-SURFACE IRRIGATION TECHNIQUE

A simple, efficient, and economic way to provide localized sub-surface irrigation known as "pitcher irrigation" was developed at the Central Soil Salinity Research Institute, Karnal, India (Mandal, 1974). In this technique, a baked earthen pot called as pitcher is used to provide and distribute water in the rootzone.

The pitcher is buried in the soil and filled with water. Through the pores, water oozes out and wets the soil in the vicinity of the pitcher, thus making it fit for sowing seeds and planting seedlings. Water is filled in the pitcher manually, through a hose or by buckets, once or twice a week depending upon the water depletion rate, which in turn depends upon the type of crop, stage of growth, and the climatic conditions. Fertilisers and insecticides are also added in the pitcher itself for uniform application. Various aspects of pitcher irrigation are discussed below.

Functioning of Pitcher Irrigation

Water discharge rates

The amount of water that can flow out though the walls of the pitcher, i.e.,

discharge per unit surface area per unit time, depends upon:

(i) The waterhead at a point: This is maximum at the bottom and thus move water flows through the bottom surface.

(ii) The apertures and intensity of pores in the earthen wall: Normally, the pores are quite small and thus the discharge is slow but enough to meet the needs of the plants. Within a reasonable limit, one can increase or decrease the size of apertures by varying the ratio of clay and sand, which are the basic materials used for making the pitchers.

(iii) Evapotranspiration need of the plant: It is greater during the day than at night. The water flow is affected by the suction created by the evapotranspiration of the plant.

iv) Texture of the soil: The discharge is usually greater in a light-textured soil than in a heavy textured soil.

Wetting Front

When the pitcher is filled, water starts coming out through the pores and wetting front advances both horizontally and vertically. However, the horizontal movement is restricted and can be predicted by the following power curve, expressed as:

L = at b

Where L = length of advance from the centre of the pitcher,

t = time in days, and

a,b = are the empirical constants, the value of which depends upon soil type, evaporative demand, discharge of the distributor, and the salinity of the irrigation water. During the winter, value of a was 16.0 and for b for a 0.25. sandy loam soil. Generally, for a pitcher of 30 cm diameter and 10 litres water holding capcity, wetted surface area is around 0.7 m^2 It is sufficient to grow at least four plants of any vegetable crop.

Moisture Content of the Soil

The moisture content of the soil around the pitcher varies from a maximum of 20 per cent near the pitcher wall to normal levels of dry soil, i.e. 5 to 10 per cent, at the end of the wetting front. The maximum moisture is nearly equal to field capacity of the soil and thus is most favourable for the growth of plants. Due to limited hydraulic conductivity of the pitcher wall, moisture levels above these are rarely obtained, except at night when, because of low evapotranspiration demand, the moisture percentage may slightly increase and encourage deep percolation losses. However, these losses can be minimized by compacting the soil or putting a plastic sheet just below the pitcher.

Salinity Development in the Soil

The pitcher irrigation method being essentially a sub-surface irrigation system, the salts accumulate at the soil surface and at the boundaries of the wetted front, leaving the rootzone in equilibrium with the salinity of the irrigation water used in the pitcher.

Life of the Pitcher

Baked clay pitchers can be continuously used for 3 to 6 years. Clogging of pores due to preceipitation of salts, as with the dripper, is a remote possibility when the pitcher is kept wet continuously. Clogging of pores from outside due to salts from the soil is limited because the flux is always continuous and occurs outwards, i.e. away from the wall of the pitcher. The plant roots can neither enter the pitcher by penetrating through the pores nor seal the surface from outside. They also do not exert enough pressure to break the wall of the pitcher. To prolong the life of the pitcher, the following precautions should be taken:

(i) While not irrigating, keep the mouth of the pitcher covered. This will minimize the loss of water as well as prevent sunlight from entering the pitcher and minimize algal information and growth.

(ii) Only clean water should be used for filing the pitcher. Muddy runoff from rainfall should be used after it is passed it through a sand filter.

(iii) Before storing the pitchers, wash them with good quality water so as to remove the salts and to prevent their precipitation and subsequent clogging of the pores on drying.

Applicability to Other Situations

Pitcher irrigation is a most effective and low-cost technology for using a limited supply of good quality water and also for using saline water resources, it is most suited to the following situations:

(i) Where availability of good quality water is scarce.

(ii) Undulating areas where soils are difficult to level for uniform application of irrigation waters.

(iii) Light soils with high infiltration rate, which cannot be efficiently irrigated by surface method.

(iv) Areas with highly saline water sources, which cannot normally be used through surface method of irrigation.

Though this technique is costlier than surface irrigation system, it is worth the additional cost in view of the easy labour availability in many developing countries and the saving of water it provides.

REED-BED TECHNOLOGY TO TREAT WASTE WATER

The reed bed system is a self contained, artificially engineered, wetland ecosystem for treatment and recycling of sullage, sewage and industrial effluent.

It is designed to optimize the microbiological, chemical and physical processes naturally occurring in the reed bed. Wetland plants, such as reeds, transfer atmospheric oxygen down through their roots in order to survive in waterlogged conditions.

This creates both aerobic and anaerobic soil conditions, allowing a rich microbial diversity comprising of bacteria and fungi use organic pollutants as a food source, breaking down a wide range of organic chemical products.

So, chemical are not simply stored in the reed bed and are actually degraded into harmless smaller components. Other contaminants comprising of heavier metals are transformed from a toxic, mobile state into fixed state by complex chemical reactions in the soil. Also adsorption capacity of soil acts as a buffer for peak or shock effluent loads.

The complexity of microbial life and powerful reactions within the root zone of the soil based reed bed system results in extraordinary water cleaning or treatment capacity.

Components of reed bed system

Resembling a beautiful garden of flowering wetland plants, the reed bed system comprises of the following components;

· A sedimentation tank

· A reed bed consisting of

- Filling media
- Acclimated, aerobic, anaerobic and facultative bacteria & fungi
- Selected indigenous wetland plants (phragmite species)

Combing mechanical filtering, chemical precipitation and biological degradation in one step, the system effectively removes suspended solids and organic matter in the form of BOD and COD.

Working of Reed bed system

After primary settling in the sedimentation tank, the breakdown of contaminants is achieved by the controlled seepage of wastewater through the root zone of the selected wetland plants. Organic pollutants are broken down as a food source by microorganisms and the plants, while the other contaminants, such as metals are fixed in the humic acid (via cation exchange bonds) in the soil or mineral substrates in which the plants are rooted. During this process suspended solids, BOD and COD are reduced to an acceptable level.

This method requires a detention period of at least six hours but provision of eight hours detention time is preferable. Hence, an effective and efficient treatment of wastewater is produced within the Reed bed in a sustainable and natural manner. There is no ongoing electricity cost to run pumps and air blowers, the power source is the sun which powers the plants through the natural process of photosynthesis.

The system will not clog up over time as the design ensures that the rate of breakdown exceeds the rate of loading in the Reed bed, so the performance of a reed usually increases not decreases, over time. Biomass is removed and not permitted to clog the voids of the gravel bed. This treatment involves a careful and continuous thinning of the growth so that biological activity does not get reduced. The reed so removed can be utilized and they have considerable commercial value.

The Role of Plants

The specially selected plant species(Phragmite species e.g. Phragmite karka and Typha angustus) with in the Reed bed have the following three main functions:

- They provide the means for secondary restructuring of the soil system. The very extensive root and rhizome system creates channels for the water to pass through and also keeps the soil open through the constant growth.
- The roots of the aquatic plant introduce atmospheric oxygen down into the body of soil, facilitating the diversity of microorganisms to flourish around the plant roots. It is essential for the effective breakdown of various organic and inorganic compounds.
- The plants are capable of taking up a certain amount of nutrients from the wastewater itself.

Features of the Reed Bed System

1. Operation and maintenance free: The reed bed system has no mechanical/moving parts. There is also no wastage of energy since mechanical aeration is not needed.
2. Clean and efficient: Since all chemical, physical and biological changes occur underground, there are no strong odours or sludge formation.
3. Self-sustaining and long lasting: Once installed and become operational, the reed bed system can sustain itself for more than 40 years.
4. Adaptability: The reed bed systems are suitable for concentrations from a few mg/l 20,000 mg/l of COD and 400 mg/l of nitrogen. Their capacity varies from about $1m^3$/day to more than 10,000 m^3/day.
5. Can handle difficult effluents: As the reed bed system has diversity of microbes and the wetland and the plants, it can adapt itself to diverse types and varying shock loads of effluents, including difficult waste waters containing.
   - Organic compounds like Chlorinated hydrocarbons, dyes and sulphur containing aromatics.
   - Nitrogen compounds such as ammonia and nitrates.
   - Sulphur compounds like sulphides and sulphates
   - Heavy metals and pathogens

Applications

The reed bed system is effective in treating effluent from

- Rural areas
- Food processing industry
- Breweries
- Animal husbandry
- Sullage and sewage from residential, institutional and industrial complexes.
- Chemical industry

Benefits

- It offers a low cost, low maintenance and sustainable solution to effluent problems.
- A single reed bed system is able to treat effluents from a number of sources.
- This method requires smaller space as compared to conventional modes of wastewater treatment including trickling filters.
- The operational cost of this mode is also less, as no trained staff is required for its operation.
- The treated effluent can be safely used for irrigation and the crops grown have higher yield and do not create any health problem.

KARNAL TECHNOLOGY

The Karnal Technology involves growing tree on ridges 1m wide and 50cm high wand disposing of the untreated sewage in furrows. The amount of the sewage/ effluents to be disposed off depends upon the age, type of plants, climatic conditions, soil texture and quality of effluents. The total discharge of effluent is so regulated that it is consumed within 12-18 hours and there is no standing water left in the trenches. Through this technique, it is possible to dispose off 0.3 to 1.0 ML of effluent per day per hectare.

This technique utilizes the entire biomass as living filter for supplying nutrients to soil and plant; irrigation renovates the effluent for atmospheric re-charge and ground storage. Further, as forest plants are to be used for fuel wood, timber or pulp, there is no chance of pathogens, heavy metals and organic compounds to enter into the human food chain system, a point that is a limiting factor when vegetables or other crops are grown with sewage.

Though most of the plants are suitable for utilizing the effluents, yet, those tree species which are fast growing can transpire high amounts of water and are able to with stand high moisture content in the root environment are most suitable for such purposes. Eucalyptus is one such species, which has the capacity to transpire large amounts of water, and remains active through out the year.

Other species suitable for this purpose are poplar and leucaena. Out of these three species, eucalyptus seems to be the best choice as poplar remains dormant in winter and thus cannot bio-drain effluent during winter months. However, if area is available and the volume of effluent is small, a combination of popular and eucalyptus is the best propagation.

This technology for sewage water use is relatively cheap and no major capital is involved. The expenditure of adopting this technology involves cost of making ridges, cost of plantation and their care.

This system generates gross returns from the sale of fuel wood. The sludge accumulating in the furrows along with the decaying forest litter can be exploited as an additional source of revenue.

As the sewage water itself provides nutrients and irrigation ameliorates the sodic soil by lowering the pH, relatively unfertile wastelands can be used for this purpose. This technology is economically viable as it involves only the cost of water conveyance from source to fields for irrigation and does not require highly skilled personnel as well. This technology seems to be most appropriate and economical viable proposition for the rural areas as this technology is used to raise forestry, which would aid in re-storing environment and to generate biomass.

SOIL AND WATER CONSERVATION FOR SUSTAINABLE DEVELOPMENT

As fertile soil and good quality water have become precious natural resources, their efficient and economical use is the first and foremost action to conserve them. The practical methods for soil and water conservation can be broadly divided into two classes.

(i) Agronomic practices.

(ii) Mechanical measures.

AGRONOMIC PRACTICES FOR SOIL AND WATER CONSERVATION

By following different agronomic practices we can reduce soil erosion, increase moisture-holding capacity of soil and can minimize problems like water logging, soil salinization etc.

Following agronomic practices are performed to achieve the above-mentioned objectives.

1. CONTOUR-FARMING: Generally, as the rain falls, a lot of runoff is generated which generally leads to soils erosion on its way downward. This removes the top fertile soil along with soil nutrients and plant seeds thus leading to scanty and uneven growth of crop.

   To avoid this a simple practice of farming is done across the slope so that there are no steep slopes on the field. The ridges and furrows thus formed act as continuous barrier to the free movement of water downwards thus provides more infiltration time. Hence, the removal of soil along with nutrients is checked to a great extent leading to increment in soil fertility and crop yield.

2. MULCHING PRACTICE: Mulching is one of the simplest and beneficial practices for soil and water conservation. Mulch is simply a protective layer of material that is spread on top of the soil to prevent it from blowing and being washed away. Mulch can either be organic-such as grass clippings, straw, bark chips and similar materials or inorganic such as stones, brick chips and plastic. Conservation tillage is a common practice that creates mulch on the soil surface. It leaves the crop residue on the top of the soil as mulch. The mulching practice yields following benefits:

   (i) Protects the soil from erosion.

   (ii) Conserve moisture in soil thus saving the need for frequent irrigation.

   (iii) Reduce compaction of soil due to impact of heavy rains.

   (iv) Maintains a more even soil temperature.

(v) Prevents weed growth to check loss of soil nutrients.

3. ENHANCING THE GROWTH OF SPECIFIC CROPS: Enhancing the growth of specific crops which provide the maximum cover, reduce runoff and soil loss e.g. legume crops in general furnish a better cover and hence better protection to cultivated land against erosion than ordinary cultivated crops. These may vary from region to region depending upon topographical and climate conditions.
4. STRIP CROPPING: It is a combination of contouring and crop rotation in which alternate strips of row crops and soil conserving crops (sods) are grown either at right angles to the direction of the prevailing wind, or following the natural contours of the terrain to prevent soil erosion of the soil. When soil is detached from the row crops by the forces of wind or water, the dense soil conserving crops traps some of the soil particles and reduces wind translation and/or runoff.
5. MIXED CROPPING: In this practice two or more crops are grown in the same field at a particular time. Some of the benefits of mixed cropping are a better and continuous cover of the land, good protection against the beating action of the rain. The different crops grown in mixed cropping have their roots at different depths holding the soil more firmly thus preventing soil erosion.

MECHANICAL MEASURES FOR SOIL & WATER CONSERVATION

These measure play an important role in soil and water conservation. These are used in conjugation with agronomic practices when they alone are not much effective. The main principles of mechanical measures are:

(i) To facilitate infiltration by increasing the time of concentration.

(ii) To breakup a long slope into several short ones to decrease velocity of runoff.

Major mechanical measures to control soil and water conservation are given as under:

1. CONTOUR-BUNDING: In this practice small bunds are constructed at regular intervals across the slope of the land. This practice is very useful in arid and semi-arid areas with high infiltration and permeability rates.
2. SUBSOILING: It is basically a primary tillage operation, which consists of break opening the soil structure up to a depth of 30 to 60 cm. This practice facilitates greater infiltration rates and moisture holding capacity of the soil.
3. BASIN LISTING: In this method of soil and water conservation basins are constructed using a special implement called basin-lister. These basins are constructed across the slope. Basin listing provides maximum time to rain water for infiltration into the soil.
4. BENCH TERRACING: In this practice a series of platforms are constructed having suitable vertical drops. The range of vertical drop may vary from 2 to 6 feets depending upon prevailing conditions. The capital cost of bench-terracing is more than that of bunding initially but in longer run it proves economical.

By above-mentioned agronomic and mechanical measures, we can have efficient soil and water conservation.

WATER SHED MANAGEMENT

By definition a 'Watershed' is a drainage area (basic hydrogeological unit) on earth's surface from which runoff resulting from precipitation flows past a single point into a large stream, a river, a lake or the ocean.

The runoff generated can be collected in a pond or a reservoir by means of mechanical/structural and vegetative measures. This stored water is then supplied to fields under gravity flow for irrigation purposes. A watershed maybe of any size but it must be manageable hydrologically. It is very much 'natural' and it allows planners to focus on all the effects of downhill runoff in a given area and to plan accordingly to control or contain it.

Watershed based management has been accepted as the most rational approach today. Besides water harvesting with special reference to rainfed agricultural lands, it includes a wide range of activities including soil and water conservation, development of degraded and waste lands, afforestation and employment generation amongst rural people.

In the long run, it preserves the ecosystem, restores the environmental degradation and stablises and sustains the overall productivity of an area. The key concepts which are included in watershed management are:

- Landuse planning and utilization based on potential.
- Rain water harvesting
- Maintaining good vegetative cover over the ground of minimize erosion of soil and water and maximize water retention in the soil.
- Draining out excess water to pond/reservoir for future use.
- Developing relevant soil and water conservation structures to control erosion.
- Maximizing productivity per unit area, per unit time and per unit of water and increasing cropping intensity.
- Utilization of marginal lands to its maximum potential use.
- Maximizing and stabilizing income of the people of the areas through integrated crop-livestock tree labour complex.
- Developing rural energy management systems, infrastructural facility and improving health, education and other socio-economic conditions of local residents.
- Finally, restoration of the ecosystem towards achieving sustainability

ROOF-TOP RAIN WATER HARVESTING FOR AUGMENTING GROUNDWATER STORAGE

- Conservation of roof top rain water in urban areas and utilizing it to augment ground water storage by artificial recharge
- Requires connecting the outlet pipe from roof top to divert collecting water to existing well/tubewell/borewell or a specially designed well.
- Urban housing complexes and industrial buildings having large roof area or a group of residential buildings can be utilized for this purpose.
- Conservation of roof top rain water in urban areas and utilizing it to augment ground water storage by artificial recharge.
- Requires connecting the outlet pipe from roof top to divert collecting water to existing well/tubewell/borewell or a specially designed well.
- Urban housing complexes and industrial buildings having large roof area or a group of residential buildings can be utilized for this purpose.
- Why it is required.
- To augment ground water storage and control decline of water levels.
- To improve ground water quality
- To reduce surface runoff which chokes the storm drains and to avoid flooding of drains
- To reduce soil erosion.
- To increase the availability of water

How to do it

The rain water from roof top is collected and the flow is diverted to the following water recharge structures:

- Abandoned dug well.
- Abandoned running hand pump.

- Recharge pit.
- Recharge trench.
- Gravity head recharge well.
- Recharge shaft.
- Defunct borewell.

Advantages of Roof Top Rain Water Harvesting

- This is an appropriate option for augmenting ground water storage in urban areas, where natural recharge has been considerably reduced due to increased urban activities and not much land is available for implementing any other artificial recharge measures.
- Monsoon run-off which otherwise flows into sewers and storm drains and is wasted is harvested and utilized.
- Rainwater is bacteriologically safe, free from organic matter and is soft in nature.
- It helps in reducing the flood hazard.
- It improves the quality of ground water through dilution, specially for fluoride & nitrate.
- Rain water may be harnessed at place of need and may be utilized at the time of need.
- The structures required for harvesting rain water are simple, economical and eco-friendly.

GROUND WATER REPLENISHMENT

The main source of groundwater is precipitation. The rate of replenishment of the groundwater varies with the pattern of precipitation, surface runoff along with the intrinsic permeability of the soil and other materials through which the water must percolate to reach the zone of saturation.

Now a days due to overexploitation of ground water for irrigation purposes, the water table is falling by about 20-25 cm per year in regions of Punjab, Haryana and Gujrat. Therefore, there is urgent need for ground water recharging.

Method of Artificial recharge

Ground water recharge may be increased by soil conservation measures and artificial recharge procedures. The amount of water infiltrated into the soil varies with the condition of soil surface and the moisture content of the soil at the time of rainfall.

The total amount of water infiltered depends on the infiltration opportunity time, which depends mainly on the slope of the land and the field structures like contour bunds, terraces and other structures, which tend to hold the runoff water over long periods on the land surface.

Water spreading

This method is similar to that of surface irrigation, with the difference that in recharge operation the water is allowed to stand on the field for a much longer period. This method may be further classified into (i) basin method, and (ii) furrow or ditch method depending upon the topographical conditions of the soil.

Ground water recharge through wells and pits

Depending upon the depth of the impermeable layer, ground water recharge can be done by large pits or wells, former being used where the impermeable layer is encountered at large depths. In this method water is allowed to infiltrate into the soil from large pits filled with water. Sometimes water is recharged into the soil directly through the tube-wells. The excess surface water is directly feed into ground through wells/ tube-wells. Mostly this method of recharge used where sufficient land is not available for water spreading. The volume of water recharged by the well or pit method is, however, low and the method is not practised where surface methods are feasible.

IRRIGATION WATER CONSERVATION

As irrigation accounts for 92% of water consumption, therefore proper attention is required to conserve this water. To meet this requirement we have following two methods:

1. Sprinkler system.

2. Drip irrigation.
3. On field Practices

Sprinkler system

Ground water recharge may in increased by soil conservation measures and artificial recharge procedures. The amount of water infiltrated into the soil varies with the condition of soil surface and the moisture content of the soil at the time of rainfall. The total amount of water infiltered depends on the infiltration opportunity time, which depends mainly on the slope of the land and the field structures like contour bunds, terraces and other structures, which tend to hold the runoff water over long periods on the land surface.

Adaptability of Sprinkler system:

- Sprinkler irrigation can be used for almost all crops (except rice and jute) and on most soils. It is however, not usually suitable in very fine textured soils (heavy clay soils) where the infilteration rates are less than about 4mm per hour.
- The flexibility of the sprinkler equipment and its efficient control of water application make this method adaptable to most topographic conditions without extensive land preparation.
- The amount of water can be controlled to meet crop needs, and light application can be made efficiently on seedling and young plants.
- Soluble fertilizers, herbicides and fungicides can be applied in irrigation water economically with little extra equipment.
- Sprinkler irrigation is becoming increasingly popular in India in regions of water scarcity where available water is insufficient to irrigate the command area by surface irrigation.
- Sprinkler are also being increasingly used for irrigating high valued plantation crops like tea, coffee, cardamom and orchards.

COST CONCEPT FOR SPRINKLER SYSTEM (2001)

Sprinkler System

Cost for constructing well and installing motor. = Rs. 1.00 lac.

Cost for electricity connection. = Rs. 0.10 lac.

Sprinkler System Set

(a) Cost of Aluminium set. = Rs. 0.50 lac.

(b) Cost of PVC set. = Rs. 0.30 lac.

Each Set Contains

(a) 40 Main/Lateral lines (30 pipes for supply, 10 spraying).

(b) 10 nozzle sets.

(c) 1 T-Section.

(d) 2 Bands.

(e) 1 End joint.

Area covered = 15 feet on both sides of the lateral lines.

Total time to cover an area of one acre = 12 hours excluding shifting time.

With each set one can cover maximum area = 10 acres.

Source: Irrigation Theory and Practices by A.M. Michael

Drip Irrigation

Drip or trickle irrigation is one of the latest methods of irrigation which is becoming increasingly popular in areas with water scarcity and salt problems. It is a method of watering plants frequently and with a volume of water approaching the consumptive use of plants, thereby minimizing such conventional losses as deep percolation, runoff and soil water evaporation. In this method irrigation is accomplished by using small diameter plastic lateral lines with devices called "emitters" or "drippers" at selected spacing to deliver water to the soil surface near the base of the plants. The system applies water slowly to keep the soil moisture within the desired range for plant growth.

In comparison to surface irrigation methods, drip irrigation can achieve 90 percent or more application efficiency.

Adaptability of drip irrigation system

- Crops like grapes, sugarcane, papayas, banana, guava and most other types of fruit trees and vegetables have been found to respond well to drip irrigation.
- There is considerable saving in water by adopting this method since the water could be applied almost precisely to the root zone and there is no need to wet the entire area between tree crops.
- Amount of water supplied can be varied depending on the stage of growth of the plant.
- The method reduces salt concentration in the root zone when irrigated with poor quality ground water.
- Like the sprinkler method, drip irrigation permits the application of fertilizers through the system.
- Moreover, the drip method of irrigation operates on much lower line pressure, thus providing a saving in energy requirement.

COST CONCEPT FOR DRIP SYSTEM (2001)

Drip System

Cost for constructing well and tank.	= Rs. 1.00 lac.
Cost of electricity connection.	= Rs. 0.10 lac.
Cost of Drip Set containing Main/Lateral. lines, emitters, joints, etc.	= Rs. 0.60 lac.

Time taken to cover a given area depends upon the cropping intensity and soil characteristics. Generally drip system is used in orchards.

Source: Irrigation Theory and Practices by A.M. Michael.

On field Practices

- Agronomic practices like off-season tillage (prior to pre-monsoon showers) conserve soil moisture. Moisture penetration to a depth of 90 cm is achieved if the land is ploughed to a depth of 30 cm. Other practices like early sowing of seeds, moderate use of fertilisers, weeding, pest and disease control and timely harvesting increase the yield inspite of limited moisture in the soil.

- Mulching the soil with organic residue conserves soil moistures.
- Terrace cultivation of hilly slopes prevents water run off.
- Contour ploughing and planting of grasses and trees check run-off water and increase the soil's capacity to retain moisture.
- Green manuring (incorporation of fresh green leaves into the soil) and crop-rotation (cultivating different crops in rotation depending upon the soil and climate, e.g. cereals followed by legumes) conserve soil moisture.

ON-FARM WATER MANAGEMENT

Increased supply of water through irrigation canals and groundwater exploitation through extensive pumping throughout the year combined with poor, drainage is causing water logging & salinity problems in large parts of the Punjab and Haryana.

Both these problems have converted a large fertile area into barren lands, as vegetation is unable to flourish in such soils. It has been shown experimentally that channel losses in water courses are 30-50% rather than 11-20% as assumed previously. Generally on farm water management programme comprises the following three components.

1. Water Course Improvement.
2. Precision Land Levelling.
3. Water Management Extension to farmers.

(A) Water Course Improvement: It includes

- Redesign and construction of watercourse to eliminate unnecessary twists & turns.
- Installation of concrete division boxes and turnout structures to provide better water control.
- Lining of channels, to reduce water losses.

(B) Precision Land Levelling: By obtaining précised land leveling through different agricultural practices, improved irrigation efficiencies can be achieved reducing water losses.

(C) Water Management Extension to farmers: By providing farmers with latest techniques of irrigation and farm management efficient farm production can be achieved.

Farmer's participation is an integral part of on farm water management due to following reasons:

- Farmers are expected to contribute their labour, thus reducing the cost significantly.
- In working on their own watercourse, it is expected that farmers would develop a sense of ownership and responsibility and would maintain it well.
- Working with skilled personnel would provide training for the farmers about water losses and means of preventing them.

Benefits of the On-Farm Water Management

- Reduction of water losses.
- Slowdown of water table fluctuations.
- Dissemination of knowledge concerning water losses among farmers community.
- Contribution to labour by farmer's community.
- Establishment of water user associations charged with operation and maintenance of water courses.
- Extension service for farmers describing their rights and duties in water sector.

Problems associated with On-Farm Water Management

- Lack of co-ordination among farmers due to social divisions.
- Non-provision of incentives to farmers.
- Lack of financial support.

- Lack of coordination between government and farmers.
- Lesser power in hands of water user associations.

DUCKWEED TECHNOLOGY FOR RURAL WASTEWATER TREATMENT

With the introduction of public water supply in rural areas and as a result of domestic activities a lot of sullage water is generated. This causes serious health hazards due to non-availability of proper drainage system. This wastewater discharged to ponds fouls the environment and results in unsanitary conditions, thereby jeopardizing the community health.

The available technologies for the treatment of wastewater involve high capital as well as operational and maintenance costs and also offer no economic returns, hence, there is an urgent need for alternate technological options.

The duckweed based waste water treatment system in conjunction with pisciculture is one such technology that has the potential of offering effective waster water treatment besides providing economic returns as well as generating employment opportunities in the rural areas.

The duckweed is a small free floating aquatic plant often seen growing in thick mats on nutrient rich fresh and brackish waters in tropical and semi tropical climate all over the world.

It belongs to botanical family Lamnaceae consisting of four genera namely; Lemma, Spirodela, Wolffia and Wolffiella of which first three genera are commonly found in India. The duckweed has the ability to bio accumulate up to 99% of the nutrients, dissolved solids and even heavy and toxic elements of wastewater up to certain extent.

These are permanently removed from wastewater as plants are harvested. Hence, it reduces Biochemical Oxygen Demand (BOD), Chemical Oxygen Demand (COD), suspended solids, bacterial and helminthic pathogens, some organic compounds, ions of potassium, nitrogen, phosphate and even heavy metals of wastewater to a level quite safe for disposal.

Further, the duckweed plants can double their mass in about two to three days under ideal conditions of nutrient availability, sunlight and temperature. The average daily yield of fresh duckweed from a pond of one hectare is about 0.5-1.5 tons. It is highly nutritious feed due to high content of proteins, fiber content and vitamins A & C.

The prolific growth of duckweed can be harvested daily from the duckweed pond and can be utilized as fish feed. Hence, not only the duckweed treated wastewater can be used for pisciculture but also the harvested duckweed can serve as complete feed for fish. Also interestingly, the duckweed species grow faster in warm weather, when fish need more feed and more slowly in cold weather when the fish also do not require as much feed.

Methodology

- To start a duckweed based sullage treatment unit in village pond, it is desirable to take up the duckweed project in villages where two separate ponds are available near by. The existing ponds are deepened, cut and dressed as per requirement. However, if the two separate ponds do not exist then a pond with a size of one hectare or more is partitioned into two equal parts (half hectare or more each) by construction of earthen bounds. Out of these two ponds, one is used for duckweed cultivation and other for pisciculture. For the treatment of sullage it is important to give it proper retention time (minimum 12-15 days) in the duckweed pond. Therefore, the depth of the duckweed pond from the invert level of incoming drain is maintained to give the sullage the minimum required retention time for treatment.
- The minimum viable size of pond for implementation of this technology is one hectare (to be partitioned into two equal halves) as quantity of sullage discharged by village having population of 4000-5000 can successfully meet the nutritional requirements of duckweed in half-hectare pond and can thus be integrated with pisciculture.

- Most of the village ponds are located in low-lying area and in rainy season, the area around the pond gets flooded. For successful execution of the project, bunds of the pond are raised so that the pond does not get flooded during the rains. Alternatively, bypass arrangement on the inlet drains along with a shutter is provided to check the flow of excess water into the ponds.
- The over flow pipes encased with wire mesh are embedded in the raised earthen bunds of fishpond to allow over flow of water in case of heavy rains. The proper slopes and compaction of the bunds is very important for the development of ponds. The minimum slope required for preparation of the bund is kept 2:1 (2 horizontal and 1 Vertical)
- The sullage brought by the main drain of the village after passing through the inlet screen (Where all the suspended debris are removed) is allowed to enter in the duckweed pond. About 100 Kg of duckweed is inoculated in duckweed pond. A floating bamboo grid (details given below) is used to check the drifting of duckweed towards one side by wind in the duckweed pond.
- For pisciculture, six varieties of fish viz. Catla, Rohu, Grass Carp, Common carp, Silver carp and Mrigal (suitable for flood water cultivation) cultivated in a separate fresh water village ponds. The sullage, after duckweed treatment is fit for pisciculture using the said varieties.
- The treated sullage from the duckweed pond is allowed to flow to the pisciculture pond through culvert pipes encased with fine mesh as and when required and excess water if any is used for irrigation purposes.
- The duckweed grown in the sullage pond is fed to fish through open bottom fish feeding stations, which comprise 4m X 4m floating PVC grid. Four fish feeding stations are required in a fishpond of half-hectare size.

Sites adopted under Duckweed project in various Districts of Punjab

S.No	Village	Block	District	Estimated Cost (R s. in lakhs)	Remarks
1	Sanghol	Khumanoo	Fatehgarh Sahib	2.20	Project commissioned in 2001
2	Chanarthal Kalan	Sirhind	-do-	3.14	-do-
3	Sandhua	Chamkaur Sahib	Ropar	3.19	Project commissioned in 2003
4	Ajnoha	Mahilpur	Hoshiarpur	3.58	Work under progress
5	Manak Dheri	Hoshiarpur	-do-	2.82	-do-
6	Villa Teja	Fatehgarh Churian	Gurdaspur	3.59	-do-
7	Sahowal	Gurdaspur	-do-	3.94	-do-
8	Takhan wadh	Moga-1	Moga	2.10	-do-
9	Kot-Ise-Khan	Kot-Ise-Khan	Moga	3.80	-do-
10	Multania	Bathinda	Bathinda	3.60	-do-
11	Marhi	Nathana	-do-	2.92	-do-
12	Dhelwa	-do-	-do-	3.65	-do-
13	Burj Gill	Bathinda	-do-	5.52	-do-
14	Gill Khurd	Rampura	-do-	3.32	-do-
15	Bhai Bhaktaur	Mour	-do-	4.12	-do-
16	Lakha Singh Wala	Amioh	Fatehgarh Sahib	Nil	Project proposal for Rs. 4.70 lacs submitted to DRDP
17	Chunni Kalan	Khera	Fatehgarh Sahib	Nil	Project proposal for Rs. 7.83 lacs submitted to DRDP
18	Chunni Khurd	-do-	-do-	-do-	Project proposal for Rs. 5.00 lacs submitted to DRDP
19	Saleempura	-do-	-do-	Nil	Project proposal for Rs. 5.72 lacs submitted to Distt Adm. Ropar

CONCLUSION

The duckweed, that is daily harvested from the half hectare duckweed pond, being highly nutritious can be utilized to feed the fish cultivated in half hectare pisciculture pond, thus saving the costs involved in purchase of fish feed. The income generation from pisciculture pond will motivate the beneficiary to adopt the technology for wastewater treatment. Besides effective sullage water treatment, the project will also generate self-employment as the Panchayats can give the ponds on lease to the local residents of the village.

7

Environment and Natural Resources Development

The present chapter identifies the economic crisis that engulfed the region in 1997 as well as the uncertainties related to proven ore reserves that has had a major impact on investment flows during the last two or three years. Regulations imposed by the growing framework of international environmental laws and guidelines on environmental value and life-cycle analysis in mining have exacerbated the problem. A tendency to use these restrictions to create discriminatory trade barriers in the mineral sector could affect the supply and demand scenarios. The document also discusses the emerging socio-economic problems resulting in the centralized control and management of mineral resources development as well as the remedial measures that could be taken. The importance of fostering partnerships among stakeholders has also been emphasized.

The second part of the document analyses the issues related to the management of land resources vis-à-vis the ever-growing population. As envisaged in Agenda 21, emphasis is being placed on the application of knowledge and information on land and its resources aimed at sustainable development. In that regard, the Forum on Urban Geology in Asia and the Pacific has been catalytic in raising the awareness of local authorities and geoscience institutions as well as inducing national and local governments to promote and support the incorporation of relevant geological information in the urban planning process.

The role of geological survey departments in land resources management is expanding as a result of newly emerging potential uses of the subsurface as well as constraints posed by geological conditions that must be respected in order to ensure sustainable urban-industrial development.

INTRODUCTION

The Asian financial crisis that engulfed the region in 1997 drastically reduced the volume of trade and the prices of basic commodities, including minerals. Furthermore, investment flows to the region also fell, significantly undermining its economic progress. Consequently in 1998, developing economies overall achieved a zero growth rate.

However, the above situation showed a turnaround in 1999 and it is expected that by the end of the year 2000 there will be a steady recovery of stock markets and currencies, moderate inflation, a downward trend in interest rates, expansionary fiscal policies and increased foreign exchange reserves, resulting in an increased volume of trade and industrial output that will trigger the reactivation of the mineral industry.

Emerging issues that would impede sustainable mineral resources development have been discussed under five subtopics:

(a) emerging issues with regard to foreign investment flows in the mineral sector;

(b) international environmental law constraints and the impact on recycling for secondary metals;

(c) environmental costs related to life-cycle analysis from mining to finished mineral products;

(d) decentralization of mineral resources development from central government to provinces and the sharing of resource rent; and

(e) emerging socio-economic issues and strengthening the partnerships of all stakeholders.

An overview of the above issues in the Asian and Pacific region was carried out with the objective of stimulating discussion on plans and action to be undertaken to address these issues in the countries concerned. The ever-growing world population renders the management of increasingly scarce land resources more and more crucial. Agenda 21 of the United Nations Conference on Environment and Development has anticipated this need and urged member countries to use all the knowledge and information available on land and its resources to alleviate wrong and wasteful types of land use and ensure that its development is sustainable. In the present document, the broader, integrative view of land resources as defined in Agenda 21 is used.(1) In land-use planning, there are both resource and hazard aspects to consider when determining whether or not a particular stretch of land is suitable for a certain type of use.

Land is valued for the space it provides, in other words, for its real-estate value, which on occasion exceeds the estimated value of the mineral resources or natural construction materials that it contains. In some cases, the environmental value of land may exceed the value of its mineral resources as well. Whatever the resource value of land, geology-related natural hazards must also be considered and may further limit the options of land-use planners. Some geology-related hazards and countermeasures are highlighted in the document.

National geological survey departments often already possess most of the relevant knowledge, but do not always report the information in a suitable format, such as easy-to-read thematic maps. Moreover, geological departments are not always in regular contact with planning institutions and decision makers. Such interdepartmental communication is crucial to avoid the adverse impacts of ill-informed planning, examples of which are touched upon later.

To enhance the presentation and communication skills of geoscientists while raising the awareness levels of decision makers, densely populated urban areas should be targeted. The Forum on Urban Geology in Asia and the Pacific proved instrumental not only in raising the awareness levels of local authorities and geoscience institutions, but also in inducing national and local governments to commit their human and financial resources to ensure a regular supply of geological information for incorporation in the urban planning process.

Countries that have participated early in this particular ESCAP programme have made markedly faster progress in integrating geology into urban planning than those that have not. It may be beneficial to redirect ongoing activities to countries that have had less opportunity to participate, while making use of the expertise already built up in those that have had a head start, supplemented by assistance from supporting countries outside the region.

This approach should include technical cooperation among developing countries, and may constitute a model for cooperation between subregions with clear socio-economic benefits, both immediate and long-term.

EMERGING ISSUES IN MINERAL SUPPLY FOR SUSTAINABLE ECONOMIC DEVELOPMENT

Emerging Issues with Regard to Foreign Investment Flows in the Mineral Sector

In the review of mineral policy and mining legislation carried out during the period 1985-1990 in most developing countries of the region with mineral endowments, the international mining companies showed increased interest in investing in mineral exploration. As a result, a significant volume of capital flowed into the region, particularly to China, India, Indonesia, Pakistan, the Philippines and Viet Nam. However, in 1997, this flow slowed and even stopped in some areas owing to the financial crisis and the unfolding of a scandal by a junior exploration company in Indonesia. At the same time, a downward trend in metal prices, in particular of gold, has further aggravated the situation.

Early in 1999, however, there were indications that investment flows were improving and that most countries were bringing in new legislation to strengthen investor confidence by adopting new rules for reporting ore reserves (the United Nations Framework Classification for Reserves/Resources: Solid Fuels and Mineral Commodities), registering junior mining companies in national stock exchanges to obtain risk capital for exploration, and independent assessment of exploration results.

The overall foreign and local investment climate for the mineral sector during the first decade of the new millennium may be considered encouraging, with countries such as China, India, Papua New Guinea, the Philippines and Thailand competing for such capital flows.

International Environmental Law Constraints and the Impact on Recycling for Secondary Metals

The recovery of secondary metals is an important trade worldwide, with an estimated annual turnover of over US$ 37 billion. This is very significant for major importers and exporters such as Japan and the Republic of Korea. This trade has many advantages, such as a decrease in demand for primary ores/minerals, the elimination of waste on land, the lowering of fixed capital investment on mines and refineries, and the provision of raw material for the further processing of metals. However, environmentalists and some government decision makers consider secondary metals to be "hazardous waste" and a threat to the environment.

During the 1980s, transboundary trade in hazardous waste and toxic waste dumping in developing countries by the developed countries created serious and national concern, which triggered the enactment of multilateral environmental agreements and national laws to address the issue. Such agreements may be categorized as those promoting transboundary trade; those that totally ban the trade; and those that merely seek to regulate the trade, such as the Basel Convention on the Control of Transboundary Movements of Hazardous Wastes and Their Disposal adopted in 1989. At present, 129 countries, including European Union members, are parties to the Convention; Brunei Darussalam and the United States of America are not parties to the Convention.

In 1995, the Basel Conference of Parties voted to create an Annex VII list of 30 (members of the Organisation for Economic Cooperation and Development and the European Union) developed countries and ban all shipments for the disposal of hazardous waste from them to non-Annex VII parties immediately, and by the end of 1997 for recovery. However, the

"Basel Ban" is not effective as it still needs ratification by some 62 countries. To date, 12 countries (Nordic countries, the European Union and the United Kingdom of Great Britain and Northern Ireland) have ratified the ban. It is opposed by Australia, Canada, Japan, New Zealand and the United States of America; none of the countries in the Asian and Pacific region, except for Sri Lanka, has ratified the ban. However, in 1998, the European Union unilaterally put the ban into legal effect for its member countries.

In the light of the foregoing, the major issues facing the recycling of secondary metals are:

(a) an inadequate definition of hazardous waste;

(b) the absence of a liability and compensation protocol;

(c) trade expansion through article II of the Basel Convention; and

(d) World Trade Organization (WTO) free trade stipulations.

In 1998, in order to address the issue of the inadequate definition of hazardous waste, the Basel Conference of Parties created two lists of hazardous waste: List A or Annex VIII (toxic) and List B or Annex IX (non-toxic), to which secondary metals belong. The remaining issues are being negotiated, and no final decisions have yet been taken by the Conference of Parties. As the "Basel Ban" may be viewed as a discriminatory trade barrier, it may be inconsistent with the General Agreement on Tariffs and Trade/WTO free trade rulings.

Environmental Costs Related to Life-cycle Analysis of Mineral Resources Development from Mining to Finished Mineral Products

Life-cycle, or "cradle-to-grave", analysis is an important tool for environmental management in the mineral sector, which can be applied to mining, mineral processing and smelting, and the manufacture of mineral-based products.

Some countries in the region are striving to standardize the methodology of life-cycle analysis by adopting the ISO 14040 series, according to which the analysis should consist of:

(a) goal and scope of definition;

(b) inventory analysis;

(c) impact assessment; and

(d) interpretation.

The major barriers to the development of life-cycle analysis in the developing countries of Asia and the Pacific are the lack of proper understanding of its importance, the lack of cooperation in data-sharing, and financial constraints.

Factors to improve the development and application of such analysis include government support; standardized methodology for life-cycle analysis; the availability of expertise and guidance; and international cooperation for data management and regional impact assessment models.

Decentralization of Mineral Resources Development from Central Government to Provinces and the Sharing of Resource Rent

In most countries of the ESCAP region, major development projects concerning metallic and non-metallic minerals do not bring any economic benefits to the local people. Resistance to such resource development intensified with the occurrence of the Asian economic crisis, which reduced resource allocations to provincial authorities by national governments.

Some governments, particularly those of Indonesia and the Philippines, have felt that there is an urgent need to address decentralization and revenue-sharing issues, and have undertaken studies for sharing resource rentals such as royalties, land rent payment for the use of infrastructure and activities through their national geological survey and mines bureaux.

In the Philippines, the legal framework for decentralization and resource rent revenue-sharing is set forth in the provisions of the Local Government Code of 1991.

Some of the problems of decentralization include unequal allocation of resources between cities/barangays and provinces/municipalities; delays in provincial government deliberations; increased involvement of non-governmental organizations (NGOs) and private sector representatives; and the inability of the provincial authorities to be actively involved in environmental management.

In Indonesia, the government enacted two major laws in 1999, the Law of Regional Autonomy and the Law on Intergovernmental Fiscal Relations. The first law grants extensive authority to 26 provinces in all matters including small-scale mining except defence, foreign, judicial, monetary and religious affairs. The second law, to be enforced by the Ministry of Finance, will provide a specific share of revenues from oil, gas and mining development to mineral resource-rich provinces.

Emerging Socio-economic Issues in Mineral Resources Development and Strengthening the Partnerships of all Stakeholders

The globalization of the mineral industry during the last decade has created some socio-economic issues that called for the strengthening of partnerships between all stakeholders, including the government, the private sector, host communities, miners, lending agencies and NGOs

An upsurge of illicit small and artisanal mining operations in some countries has partly resulted from the recent economic downturn. Such operations focus mainly on precious metals, such as gold. As mining is often the only source of income for people who live below the poverty line, illicit mining operations have to be regularized in remote areas where host communities are involved.

Several countries are now planning to enact new laws and regulations for small-scale and artisanal mining operations, while assistance is being sought from United Nations agencies and donor countries to combat mercury and arsenic contamination arising from illicit gold mining.

To address the socio-economic problems related to mineral resources development, it is imperative that partnerships between all stakeholders be strengthened. This would contribute significantly towards curbing environmental degradation resulting from mining.

EMERGING ISSUES RELATED TO LAND-USE PLANNING FOR SUSTAINABLE ECONOMIC DEVELOPMENT

Emerging Importance of Geology in Sustainable Development

Most modern land-use planning aims at an optimum distribution of the population and its activities. To reach this goal, it is necessary to strive for the optimum and integrated management of natural resources, and try to adapt infrastructure and housing projects to environmental conditions. The final objective of such plans should be the long-term optimization of master plans, the protection and rehabilitation of vulnerable areas and the prevention of natural disasters and pollution.

Geology plays an increasingly important role in this. The soil and the subsurface form a physical environment of great diversity and development potential, but one that is at the same time very fragile. In this context, space and landscape, water, energy and mineral resources, the environmental-bearing capacity of towns and infrastructure, and the potential for receiving and confining human waste products are all potentially valuable resource elements. However, those resources are limited and mostly non-renewable, vulnerable and already partly degraded by human activity. It is therefore necessary to rationalize and optimize their use through the sustainable development of subsurface resources; the rational organization of space, amenities and construction; the protection and rehabilitation of vulnerable space; and the prevention of hazards affecting the subsurface.

Subsurface resources should be taken into account in any land-use development policy, regardless of whether such resources play a positive role in the economic development and satisfy a demand, or whether they place restrictions on land use. Geology, while making use of many different disciplines such as metallogeny, geothermal-energy development, hydrogeology, geochemistry, geophysics or geotechnical engineering, will be able to identify and evaluate such resources, as well their quality and lifespan. Furthermore, geology will help set guidelines for environmental preservation and for the sustainable management and protection of any renewable resources.

Geology and Groundwater Resources

Groundwater is a natural resource with a rather constant quality. Therefore, in many urban areas drinking water is mostly groundwater, which is often exploited without considering the recharge capacity of the aquifer, or the impact of such exploitation on wet-lands, river flow and ground subsidence. Aquifers can also be vulnerable to pollution, depending on the overlying ground characteristics. Groundwater resources are important not only to the economy but also in the equilibrium of ecosystems, and must therefore be considered in conjunction with surface water when drawing up land-use plans.

Many countries possess a wealth of thermal and mineral waters at sites that may have therapeutic value and attract tourists as well. In parallel with the growing demand for mineral water by urban populations, thermal water requirements grow with the influx of tourists. Such resources may dwindle and degrade rapidly, as they are highly vulnerable to pollution. Plans for the creation of spas and mineral-water exploitation must therefore be thoroughly scrutinized.

In parts of the region, notably in Bangladesh, China and India, groundwater was seen as a way out of surface-water pollution problems, which caused much suffering to their mostly rural population. The use of groundwater has indeed substantially reduced the number of deaths due to water-borne diseases. However, as is well publicized, a large number of tube wells in Bangladesh now produce water containing increasing amounts of arsenic, causing widespread health problems such as skin lesions, and eventually death, among the population. As the phenomenon only surfaced after the tube wells had been in use for years, it has been impossible to predict whether or not a certain well would be safe in the long term. An international effort would be required to determine the precise conditions under which this toxic element becomes mobile and to identify the exact geological formations in which those conditions prevail.

Only with such a predictive capability may further groundwater use for public consumption be sanctioned in vulnerable areas.

Mineral Resources, including Natural Construction Materials

In the present context, mineral resources are mentioned only as far as they affect the value of overlying land, restrict current or sequential land use, or may pollute the soil, groundwater and surface water, and the atmosphere. Thorough geological surveys are needed to determine both the economic and environmental aspects of a potential mine or a particular mineral deposit. Another problem is the sterilization of mineral resources, which can occur when engaging in construction before proper mineral resources assessment has taken place for that location.

Sterilization by overlying construction is common in the case of natural building materials, such as sand or gravel, occurring at a shallow depth. The same construction projects often import sand or gravel from far away, thus driving up the price unnecessarily. Again, a simple geological survey of the area would have resulted in large savings.

Aggregate and related materials are needed for housing and infrastructure projects, but such heavy materials, with their low initial value, have to be destined for nearby markets. The environmental impact of excavating them, usually from open quarries and sometimes from streams, and the chance of finding substitute materials, have prompted local authorities in some countries to develop their own plans for aggregate quarries, in partnership with private enterprise.

Alternative Energy Resources

Several sites in the region have been developed for the production of hydroelectric power, either based on the relief and hydrologic potential of the catchment area or on the geological (geotechnical) quality of the host rock for dams and reservoirs. Although many major hydroelectric infrastructure elements have been built, the construction of hydropower installations remains an option at the regional level.

The 1970s saw some experiments with subsurface geothermal heat-generation networks for heating entire urban districts. The technical difficulties of this process have slowed such geothermal development. However, enormous technical advances have since been made and geothermal energy is now considered a viable alternative, as testified by a resumption of activity in New Zealand and the United States of America.

Another possibility lies in hot (dry) rock in areas with a strong geothermal gradient; research into this alternative is progressing and might lead to new auxiliary energy sources. Recent work in Hawaii indicates that volcanoes may constitute an interesting energy source that merits further geological investigation.

Potential uses of Underground Properties

Considering soil and subsurface properties is not entirely new in land-use planning, but should be generally adopted to ensure better environmental protection and sustainable development. This means better use of the subsurface, which translates into preservation of landscapes and ecosystems, energy savings, human health and safety.

Underground storage of products not only brings improved security conditions but also saves space on the surface. The many rock types that exist make it possible, for example, to store natural gas in highly porous aquifer formations, to store liquids in cavities excavated in impermeable formations, and to dispose of toxic or solid waste in rock formations.

Current technology for excavating underground cavities has made feasible the construction of underground transportation routes, car parks and other installations, leading to new concepts of underground urban development. The stable temperature underground is another advantage for the storage of products.

Temporary subsurface storage of surface water is a newly emerging possibility, whether for recharging aquifers supplying domestic water or for handling excess rainwater to avoid flooding, or to avoid pollution from urban/industrial runoff.. Research into the purifying properties of certain soils and geological environments may eventually contribute to the cleaning of effluents at very low cost.

Reducing carbon dioxide (CO_2) emissions in order to control the overall levels of CO_2 in the atmosphere has become an international priority in the wake of the Kyoto Protocol to the United Nations Framework Convention on Climate Change, adopted in 1997. Recent geological studies have demonstrated the economic and technical feasibility of reducing greenhouse gas emissions into the atmosphere by the underground storage of CO_2, either in saline aquifers or in depleted oil and gas fields.

Some attributes required for a rational use of the subsurface are straightforward, such as ground suitable for building foundations. Other potentially favourable properties are only found in areas with a specific geological setting, such as formations with the properties required for confining or trapping toxic compounds. Geological surveys should be mandated to ensure that sites with such potential are identified and mapped well in advance for incorporation in land-use planning and to avoid their sterilization by other types of land use.

Geology-related Natural Hazards

Prevailing geological conditions with active crustal plate boundaries render the ESCAP region prone to various natural hazards. Prominent among those geology-related hazards are earthquakes, volcanic eruptions, landslides, tsunamis, mud or debris flows, glacial lake bursts etc. All of these are potentially catastrophic, particularly if known information about those hazards collected by the geological survey departments is heeded neither by the population nor by the authorities.

Although these natural hazards themselves cannot be made to disappear, they do not have to develop into disasters. Disasters are inevitable only when people, possessions and productive assets are put at risk in hazard zones, but those are essentially human decisions. As geology-related hazards are attributes directly linked to location, hazard mapping is feasible and should be made mandatory in densely populated areas or areas that are soon to be urbanized. Maps delineating zones having different hazard levels are essential tools in preventing disasters, by enabling the authorities to ensure that people and property are not put at risk in hazard zones.

Seismic, volcanic or landslide hazard maps are straightforward and easy for the authorities, other professionals and the public to understand. The information needed for making such maps is mostly already on file in national geological survey departments. Often all that is needed is a dedicated effort to put those data in an easily digestible form: the thematic hazard map. Geographic information systems (GIS) technology is now readily available, affordable and easier to master than ever, and has already served to mitigate disasters. GIS for making hazard maps should empower land-use planners in partnership with geoscientists. Funding GIS can be justified by highlighting the huge potential savings in human lives and assets.

Major Areas for Integrating Geosciences in Land-use Planning

The space beneath our feet is an integral part of the human environment and a foundation to most amenities and activities, but is also surprisingly fragile, and this has mostly been ignored by developers. Disregarding the vulnerability, heterogeneity and relative instability of the subsurface has costly consequences, particularly if geological resources are degraded, which may compromise sustainable development.

Groundwater pollution is well-known, but in contrast to surface water, which recovers its initial quality once the polluting source has disappeared, the regeneration of groundwater quality takes decades. Some aquifers are protected by impermeable ground that shields them from the surface. Others are exposed to contamination and adversely affected by human activity, such as agricultural use of fertilizers and pesticides; industrial use of compounds that can migrate through soil (hydrocarbons, solvents, mineral salts or heavy metals); disposal of urban or industrial waste; or poorly controlled sanitation, such as leaking sewerage systems or ill-designed septic tanks. It is therefore essential for land-use planners and decision makers to respect the restrictions imposed by the soil and subsurface characteristics reported by geologists.

Soil pollution below derelict industrial or mining sites, storage areas for toxic products (especially hydrocarbons), or abandoned landfill sites can present a risk to health, which is even more insidious as the systematic evaluation of the chemical quality of soil is not yet routine. To make amends, geological departments should conduct baseline surveys to make an inventory of the natural geochemistry of soil and subsurface as well as old pollution sites, to enable sustainable development planning that would ensure human health and safety for the next generations. The geotechnical properties of the ground directly affect the foundation costs of buildings or the excavation costs of underground amenities. Ground subsidence can be a major problem in marshy areas or in areas underlain by karst (weathered limestone forming underground cavities). Damage to structures founded on clay that is sensitive to variations in its water content can be costly indeed, especially during cyclical climate variations (wet versus dry years). Plans for land-use development should take account of this, as well as the relative instability and heterogeneity of underlying geological formations, in order to avoid such serious risks.

The problems due to ground subsidence in cities of the ESCAP region commonly result from overexploitation of groundwater resources that cannot meet the water supply demands of urban populations. The resulting shrinkage of the underlying clay is irreversible and creates a "subsidence bowl" lacking natural drainage, causing periodic flooding of urban areas in the rainy season. Differential ground subsidence causes sewerage systems to tilt in the wrong direction, thus creating blockage and frequent breakage of pipes, which releases raw sewage into the flood waters, posing a grave health risk. Central or local governments that intend to avoid such regrettably common health hazards in the future may support the creation of urban geological advisory teams with direct links to urban planning departments.

CONCLUSIONS AND RECOMMENDATIONS

The issues that have been covered, such as foreign investment flows, restrictions on recycling of metals owing to stringent international environmental laws, environmental costs related to life-cycle analysis, decentralization of mineral resources development and sharing of resource rent, would play a significant role in the future exploitation of mineral resource endowments, especially the scarce base metals in the region.

The secretariat would like to seek the advice and guidance of the Commission on the future directions that would ensure proper understanding of the issues highlighted in the present document. To this end, it is recommended that the governments in the region should actively pursue the developments related to the Basel Convention and the "Basel Ban", as well as other issues such as life-cycle analysis, and socio-economic issues that may lead to discriminatory mineral trade practices. The issues related to decentralization and the sharing of resource rent are being studied extensively and remedial measures taken in Indonesia and the Philippines. Those two models could be studied by other countries so as to evolve a common position. The ESCAP secretariat, if requested to do so, could initiate a regional study to identify an appropriate regime for such activity by the member states.

The chapter has also covered issues emerging in land-use planning for sustainable economic development, as based on the integrative view of land, which includes the resources it contains, such as soil, minerals, water and biota; underground space is emerging as an additional land resource with diverse development potential.

Direct communication between land-use planners and geoscientists is crucial in order to guarantee the incorporation of geological conditions in land-use planning for sustainable economic development. Authorities are advised to promote such direct, "horizontal" links between the two professional communities.

Support for and participation in the Forum on Urban Geology in Asia and the Pacific would be instrumental in establishing and maintaining the crucial dialogue between the planners and the geoscience community at an international level. This would guarantee the sharing of the relevant know-how accumulated in countries with more experience in the integration of geology in land-use/urban planning.

Some central and particularly local governments have started establishing their own geology-for-urban-planning programmes, either self-financed or funded from bilateral sources. Their example would be worth emulating in order to enhance the quality of urban planning and stave off the disastrous results of ill-informed decision-making. Many different potential uses of underground space have emerged. Governments would guarantee the rational use of this resource by mandating routine geological surveys to identify and map the conditions required for such novel uses, well in advance of actual development decisions.

Groundwater resources, although a commonly clean and relatively cheap water supply source, will increasingly need to be combined with surface water in order to avoid the adverse effects of groundwater overexploitation, notably the salinization of groundwater, ground subsidence and resultant urban flooding.

Members with a problem of naturally occurring arsenic in groundwater used for domestic drinking water may benefit from comparative stratigraphic studies at the international level in affected areas to identify precisely the geological formations that release the arsenic into the water supply, in order to enable authorities to recognize and avoid potentially toxic sources.

STRIVE FOR SUSTAINED SUPPLY

THE water scene in this country is profoundly disquieting. The one positive aspect that we can think of is the contribution that irrigation undoubtedly made to the Green Revolution. We can regard that as a success story, ignoring the many qualifications that need to be made. However, there is not much more to be said on the positive side. Consider the following synoptic picture:

- Urban water supply. The state has failed to assure the availability of safe drinking water to all; what is supplied is inequitably distributed, with excessive use by some and meagre supplies to others. There is an implicit subsidisation of the rich through low water rates and an inadequate coverage of the poor by the public system, forcing them to buy water at much higher rates from private sources. There is an enormous generation of waste, with only a part of it being treated, and untreated sewage and effluents are discharged into the rivers, turning them into sewers or sources of poisoned waters.

BASIC RIGHT: It is the state's duty to ensure that everybody gets enough water to meet his daily needs

- Rural areas. There is the conundrum of constant or growing numbers of uncovered villages, i.e., villages without a nearby source of safe drinking water, despite the repeated claims of achievement of targets; and the burden of bringing water from distant sources falls on women, including girl children.
- Major and medium irrigation projects. Maintenance is poor, systems are in disrepair, supplies are unreliable, and the service is unsatisfactory. There is a vicious circle of low water rates, poor revenues, inadequate provisions for operation and maintenance, poor service, and resistance to increases in rates.
- Canal-irrigated agriculture. Water-use efficiency is low, yields are poor, there are inequities between head-reach and tail-end farms and between rich and poor farmers, conditions of water-logging and salinity in many areas, and so on.
- Inter-state river water disputes. These are persistent and bitter, and tend to get enmeshed in party politics and become intractable. The constitutionally mandated conflict-resolution machinery has run into difficulties in recent years.
- Groundwater. The explosion of groundwater use from the 1980s onwards has led to dramatic increases in production, but also to alarming depletion of the aquifers as well as their contamination. Regulation seems very difficult, if not impossible. There is also the worrying emergence of fluoride and arsenic in groundwater in different parts of the country.

- Floods. There is a failure to evolve a strategy for coping with floods and minimising damage, and an absence of proper information systems and disaster-preparedness.
- Drought-prone or arid areas (in Rajasthan, Madhya Pradesh, Maharashtra, Gujarat, Karnataka, Andhra Pradesh and Tamil Nadu). There has been a failure to evolve area-specific answers to the problems of these regions.
- Pollution and contamination of water sources and systems. In effect this reduces supplies by making much of the `available' water resources unusable.
- The failures are manifest. There has been a comprehensive mismanagement of water.

Looking back at the past, we can see that our thinking about water was influenced by the following factors:

- Irrigation loomed large in our thinking, though later we began to be concerned about industrial demand as well.
- The entire thrust in policy and planning was to make more water available for use, i.e., a supply-side response to projected or imagined demand.
- The thinking was prima-rily in engineering terms.

The engineering-dominated, supply-side approach meant that the attention was focussed on what is referred to as 'water resource development'; the manner in which water was used or managed received little attention. As the driving force behind that approach, which still continues, is the expectation of a crisis, and this needs to be looked at briefly.

Averting Crisis

Fears of a water crisis arise from projections of future demand linked to projections of the population growth rate, pace of urbanisation and processes of economic 'development'. The 'demand' should not be readily accepted; it needs to be re-examined very carefully.

The availability of canal water from major/medium projects leads to the adoption of water-intensive cropping patterns, creating demands for more water for more irrigated agriculture of the same kind. This results in projections of unsustainable future demands for water and creates scenarios of scarcity. It is this kind of competitive, unsustainable water demand that makes inter-state river-water disputes difficult to resolve. It is necessary to break free of that kind of thinking. Instead of asking for more water and still more water, the approach has to be one of getting more out of available water. That may require changes in what is grown, but even more important is the efficient use of water. If substantial improvements in yields from irrigated agriculture can be brought about, the demand for water can be significantly reduced.

In rural and urban water supply, apart from improving the efficiency of the system, it is necessary to enforce economies on those that use too much water, and improve availability to groups or areas that receive too little. If this were done, it might not be necessary to raise the average `norm' which forms the basis of projections of water requirements; in fact, in some European cities, norms for public water supply are being reduced, not increased.

Reduce Waste

In industrial use, multiple recycling and re-use needs to be insisted upon, allowing minimal make-up water: we must move towards a situation in which 90 per cent of the requirement for industry would be met through recycling. That might be very difficult today, but it must be our ultimate goal.

Finally, the amount of waste that is taking place in every use needs to be tackled: first the waste must be reduced, and secondly, a part of it must be recovered for certain uses.

On the supply side, the available water resources are measured at the terminal points of the river systems. However, what is available in nature is not just the river-flow but rainfall, which is a much larger figure.

So, prima facie, there would seem to be some scope for capturing a little more of the rainfall, in which case the available water-resource number might change. Besides, the usable component of the available supplies could be increased not merely by storage behind a large dam, but also through storage in small water-harvesting structures, or through the re-charging of aquifers.

All these are matters for research, but in a study made for the UN University at Tokyo, Professors Kanchan Chopra and Biswnath Goldar of the Institute of Economic Growth had projected a figure of 'additional run-off capture' of 140 BCM.

That number may be questioned, but it suggests that this could be a significant component of water planning.

If we do all these things on the demand and supply sides, a veritable transformation of the water scene may result, and a crisis may be averted or its severity reduced. Such an approach may not eliminate the need for big projects, but it will certainly minimise that need.

Incidentally, there are only three ways in which water available for use can be augmented: rainwater harvesting, groundwater drilling and large projects (for storage, i.e., dams and reservoirs, or long-distance water transfers). Each of these would have its impact and consequences. The impact and consequences of large dams are by now fairly well known. In recent years, the reckless exploitation of groundwater and the consequent depletion and/or contamination of aquifers have begun to cause serious concern.

Rainwater harvesting has barely begun to be promoted, but some critics have already started cautioning against extensive recourse to this. Obviously, none of these possibilities on the supply side can be ruled out altogether; a wise and prudent combination of all three would need to be adopted.

ILR Project

At this point a reference must be made to the controversial 'Inter-Linking of Rivers' (ILR) Project. The supporters of the project claim that it will moderate floods, provide water to thirsty areas, make a significant addition to irrigated area, help in the greening of India, create millions of jobs, provide a substantial net generation of hydro-electric power, and so on. The critics question each one of these claims, and say that the project will have serious environmental, social and human impact and consequences; that a favourable balance between these costs and the benefits arising from the project cannot be presumed but must be established link by link; that alternative ways of meeting the country's future needs are available; and that the project is not needed at all.

It is not necessary to discuss this further here, because the present government proposes to study each link carefully and proceed cautiously and in a phased manner.

Water Markets

Another response to the projections of future water needs is that the answer may be found in water markets. Along with this goes the advocacy of privatisation of water services. This approach is strongly advocated by international financial institutions, western governments and some of our own economists. It must be noted that the argument for privatisation cannot forthwith be transferred from consumer or industrial goods to water, because the analogy is inapt: in the event of a market failure we can do without most consumer or industrial goods, or look for alternatives, but we cannot do without water and there are no substitutes for it.

Moreover, water is a basic right, and the state does have a responsibility to ensure that no one is denied this right, even if the service provision is entrusted to a private agency. Besides, the prime motive of the private corporate sector is profit, and if considerations of profitability come into conflict with other considerations, profitability will prevail; such an approach seems inappropriate in the sphere of a basic life-support resource.

It may be argued that the service can be privatised but not the resource. It is difficult to maintain that distinction. The privatisation of the water supply service may sooner or later lead to the transfer of control over the resource itself to private hands. The transfer of control structures (a dam, a barrage, a borewell or a pumping station) to a private agency, or the building of such structures by it, would give it a position of power which cannot easily be undone, and can have serious implications. It is evident that the question bristles with difficulties. Drawing the various threads together, it is clear that water is among the most important problems facing this country. A major national effort-a multi-pronged national campaign or movement on water (encompassing both the state and civil society)-is called for. Among other things such a campaign or movement will have to:

- Restrain and manage demand within availability.
- Promote efficiency and economy in water-use and resource-conservation.
- Foster a consciousness of a scarce and precious resource.
- Facilitate the prevention or quick resolution of conflicts.
- Promote rainwater-harvesting and micro-watershed development throughout the country to the maximum extent technically feasible without adverse effects.
- Subject large projects, where these are found necessary, to a stringent evaluation procedure.
- Arrest as quickly as possible the present disastrous over-exploitation of groundwater.
- Arrest and reverse the loss of good water to pollution and contamination.
- Bring about the necessary changes in water laws needed for all this.
- Bring about a major transformation in the way people think about water.

Within the space of this article it is not possible to put forward detailed and specific recommendations on these matters. What one can stress is the importance of recognising that a special, exceptional effort is called for. It is of course much easier to build a dam or drill deep for water than to undertake that kind of effort, but the latter is what is needed.

MAJOR STRATEGIES AND ACTION PLANS

Agriculture

The Union territory of Pondicherry is an agrarian state, with a large population depending on agriculture and allied activities for their livelihoods. Over the years, the Union

Territory has faced the serious problem of dwindling land availability not only due to fragmentation of holdings, but also due to population pressure and large scale industrialization and consequent urbanization. Further, agriculturally productive lands have been put to non-agricultural purposes like providing space for housing, industries, degradive activities like brick making, which makes the soil permanently non productive. There has been a constant decline in the net area cultivated due to competing demands on the in-elastic land resource by industries and other socio-economic activities. The area put undeR cultivation has been gradually declining, while the area sown more than once had been steadily increasing over a period of preceding two decades and then started declining or stabilizing in comparison with the net area cultivated. With the increase in the intensity of cropping, there has been depletion of soil resources particularly the major plant nutrients as well as micro-nutrients. To replenish these nutrient sources, huge quantities of chemical fertilizers have been used continuously for years together thereby changing the soil reaction constantly interms of soil pH, which is one of the reasons for inhibiting the survival of the micro flora and micro fauna of the soil.

There has been a drastic reduction in the area irrigated by surface water resources. Simultaneously, there has been corresponding increase in the area irrigated by ground water resources through the network of tube wells. Consequently, large quantities of soluble salts including toxic substances like sulphides, and iron compounds are brought about and deposited on the soil surface which are also responsible for inhibiting the growth of micro organisms to a larger extent.

Normally, fallowing of cultivable lands for aspecific length of period is a desirable agronomic practice in order to enable the soil to replenish its wear and tear and to recoup the plant nutrient resources for supporting crop growth in the succeeding years. The area under current fallows, other fallows and culturable wastes are gradually increasing year after year, there is no restoration of cultivable area thereby revealing the fact that these fallowing of lands are nothing but unauthorized conversion of land use from agriculture to non-agriculture uses like housing etc. With the result, it is quite obvious that the fallowing of lands in this Union Territory is not being practiced for the purpose of leaving any period of rest to the lands and the associated micro-organisms but only for other reasons and such of those lands on cultivation is being practiced are continuously put and stress and strain with out any gap.

Loss of Agriculture Land

The Union Territory has faced the serious problem of dwindling land availability not only due to fragmentation of holdings, but also due to population pressure and large scale industrialization and consequent urbanization. Further, agriculturally productive lands have been put to non-agricultural purposes like providing space for housing, industries, degrading activities like brick making etc., making the agriculture land permanently non productive. There has been a constant decline in the net area cultivated due to competing demands on the inelastic land resource by industries and other socio-economic activities.

Loss of Native Variety and Crop Pattern

Agricultural modernization, which introduced new and uniform crops into farmer's fields have seriously eroded the diversity of local varieties. This displacement has not only led to loss of diversity, but several undesirable ecological consequences. Several paddy varieties such as Seeraga samba, Kuthiraival samba and Vaigunda are now facing extinction. These varieties, if conserved, will be of immense value to Pondicherry agriculture in many ways. A concerted effort to preserve them is vital to the future.

Continuous application of chemical fertilizer, depletion of soil resources-nutrient, inhibit growth of micro-organisms.

Over Exploitation of Ground Water Resources

The requirement of water for agriculture presently being drawn from coastal aquifers, have led to a steep decline of water level in the aquifers which have caused salt water intrusion, which is non reversible. On the basis of strategies for conservation of biodiversity in agriculture sector the following actions are recommended to be taken.

The Government should enforce strictly the law so that agriculture land is not converted for non-agriculture purposes. A joint action of Department of Agriculture, Town and Country Planning and Local Administration Department is required.

During public hearings it was noted that the farmers in the urban oriented state are not aware of biodiversity issues such as cultivation of traditional varieties of crops. It is suggested that the Department of Agriculture with the help of Krishi Vigyan Kendra and Department of Education arrange field schools to address these issues. The farmer should be made aware of the hidden advantages of traditional varieties for their food and livelihood securities. A detailed action plan is being proposed.

In order to promote traditional varieties of crop and non paddy crop produce the Pondicherry Institute of hospitality and hotel management should organize food festivals using traditional varieties and popularize the value of them. This would help increasing the demand of local produce of traditional crops. The Department of Education should promote the use of some of the traditional varieties of rice by providing them in their noon-meal scheme.

A community seed bank should be set up in order to survey and document for conservation of traditional paddy varieties, ground nut, and indigenous vegetables for seed exchange and utilization among the farmers. Characterization, and multiplication of the traditional varieties can be done, involving the farmers community through the community seed bank evaluation. The community seed bank will help the farmers to conserve the traditional varieties, which will be managed by the farming community itself. Indigenous knowledge system should be catalogued with reference to conservation of bio-resources.Seed materials and incentives should be given to the farmers who are willing to cultivate the indigenous varieties of paddy and pulses with the buy back facilities.

A technical advisory board consisting of agriculture experts should be formed for developing a community Biodiversity Register for enlisting of crop species and varieties. This will help to identify geographical indicators if any, which will be having a potential for registration as an exclusive material of Union Territory of Pondicherry. Advisory board would consist of Secretary / Director, Agriculture, Dean, Pandit Jawaharlal Nehru College of Agriculture & Research Institute, Principal, Krishi Vignan Kendra, Pondicherry, and coordinators, NBSAP, Pondicherry University.

The above work will be executed by the field level officers and staff. The field level staff should be trained in biodiversity related activities, to create awareness among the farmers and the public. Every village should have the biodiversity register so as to enable the farmers to enlist the available crop species and varieties and the traditional knowledge systems should be recorded.

Biodiversity festival and seed exchange fairs should be conducted to create awareness and the importance of conservation and to share the views on traditional ways of agriculture farming system. The fairs could coincide with traditional harvest festival like Pongal, and Tami New Year. In order to promote traditional non-grain crops, more 'Farmer's Market' could be established in the Union Territory so that the farmers could get good price of their produce. PAPSCO / PASIC should coordinate such ventures. In order to promote traditional varieties of crop and non paddy crop produce, the Pondicherry Institute of hospitality and hotel management should organize food festivals using traditional varieties and popularize the value of them. This would help increasing the demand of local produce of traditional crops.

Farmers should be educated through field schools about the importance of biodiversity conservation. The impact and the risks involved by using the chemical fertilizers, and the status of the nutrient content of the soil should be made known to the farmers. A soil health card can be made available to the farmers and periodical soil examinations can be carried out and accordingly the fertilizers can be judicially used. To minimize the use of chemical fertilizers the subsidy can be totally removed. Use of bio-fertilizers, organic manure, enriched compost, green manure seeds and bio-pesticides should be encouraged by giving more subsidy.

To assess the status of microbial population in different kind of soils, surveys have to be conducted in order to channalise the limited available materials like enriched compost, green manure seeds etc., should be made available to the areas where the intensity of microbial population is comparatively less, so as to step up the same on par with other areas.

Studies have to be conducted to assess the status of humus content in soils in terms of the available organic carbon in order to give preferential allotment of the enriched compost etc., to the deficient areas.

The above surveys are required to be conducted and completed with--in a period of two years so as to make use of the findings for implementing the corrective measures within the shortest possible time span. The surveys could be done by PAJANCOA &RI with the help of soil microbiologists. Awareness campaigns confined to the localities deficient in the microbial population as well as humus content have to be conducted so as to motivate the farmers and to treat these areas on priority basis with the whole village adoption approach. To promote organic farming, farmers should be trained on vermi composting and coir pith composting. The farmers should be encouraged to do organic farming by providing compensation for losses. Necessary steps should be established to market linkages and to certify the organic produce in order to encourage the public and the farmers. A separate market should be established to popularize organic products which in turn will create awareness among public.

Pondicherry Government should encourage the use of bio-fertilizers in place of chemical fertilizers. The use of bio-fertilizers have been very successful in neighbouring Auroville. (Tamil Nadu). Auroville model may be implemented in Pondicherry. The farmers then would be able to get more money on such produce as agriculture produce without the use of fertilizer and pesticides is in great demand in the Indian and foreign market.

Apart from conducting Field school and education about the Integrated Pest Management and agriculture practices, more emphasis should be given for identification and documentation of plants with pesticidal properties and identification of pests they can be used against. A practical feasible plan for use of non-chemical pesticides for farmers should be developed. Periodical training programmes should be conducted to familiarize the use of bio-pesticides.

Implementing agency : Secretary/ Director, Agriculture Department. Extension wing and PAJANCOA & RI, Karaikal could be responsible for implementation of the above action plan after receiving funding support from the Department of Agriculture.

Education, Public Awareness and Training

The public should be made aware of the biodiversity in agriculture by organizing workshops, symposia and biodiversity festivals. This will help public to be aware of food security, nutritional security through nutritional gardens, sustainable agriculture development, farmers right, bio-prospecting / benefit sharing, geographical indicators, intellectual property rights, importance of convention on biological diversity (CBD), bio-piracy, importance of documentation of bio-resources in the era of patent disputes etc.

The science of taxonomy of field crops should be promoted by organizing short term practical courses for students, research scholars, scientists with exposure to both classical and as well as molecular taxonomy using DNA Marker based techniques. National level Traditional Science Congress with the focus on conservation of biodiversity can be conducted.

Implementing agency : Department of Agriculture in coordination with Pondicherry University and PAJANCOA &RI. with funds allocation from the Department of Agriculture and Department of Science, Technology and Environment.

Water Conservation

Depletion of ground water resources is adversely affecting crops in Pondicherry and Karaikal areas. Conservation of ground water resources for agriculture is to be done by conducting field schools in the villages in participatory mode in crop water management and creating an awareness among farmers to use water for irrigation of crops judiciously. The village ponds in the Pondichery region should be desilted for better recharge.

Additionally tube wells can be constructed in all these ponds as rain water harvesting structures to increase artificial recharge to aquifers. Appropriate schemes should be introduced with incentives for change of cropping pattern especially in summer months from paddy to low water consumptive crops to conserve utilization of ground water.

The power supply to agriculture pump sets should be regulated to 10 to 12 hours per day instead of 24 hours.

Increase allocation of funds should be made for transportation of water from tube-wells to the fields through underground pipes to save ground water from additional seepage loss in earthen canals.

The present tariff for electricity for agriculture may be increased for agriculture pump sets to imbibe economy in usage of ground water and also to achieve irrigation efficiency in the application of water to the irrigated crops.

The Government should popularize the use of sprinkler and drip irrigation system for orchard crops where in more than 50% of water conservation is possible.

Extensive education programmes on Crop-Water management methodologies should be undertaken covering the entire Union Territory of Pondicherry for paddy, sugar cane and ground nut to impart the latest available technologies to the farmers.

Construction of Check Dams

Check dams should be constructed across the Pambai river which flows through Sorapet village in Pondicherry region and in the canal which flows in Challakara, Mahe region in order to recharge the ground water. The Challakara canal should be stone walled to control soil erosion. Implementing Agency : Director, Agriculture Department. Ground Water Unit: Town and Country Planning Department, Krishi Vignan Kendra, Electricity department, and Department of Industries.

Botanical Gardens

The existing Botanical Garden in Pondicherry houses many Government Departments and has an aquarium with it with a toy train. Due to increased human activity, release of saline water from the aquarium and aging of trees, many trees are dying and are at the verge of extinction. There is a strong need of rejuvenating the existing botanical garden and also to establish atleast one more botanical garden in Karaikal.

A new botanical garden should be established in Karaikal region with many indigenous, rare, endangered, threatened plants as well as many horticultural plants and some exotic plants. The Dean, PAJANCOA &RI, Karaikal has proposed to help the Government in this direction. An alternate botanical garden can also be created as most of the trees in the existing garden are aged, preferably in a silent zone. It is suggested that Pondicherry University can undertake this project.

Miniature gardens can be created in as many educational institutions as possible to create awareness among the school children about biodiversity conservation.

Implementing Agency : Department of Agriculture in coordination with Pondicherry University and PAJANCOA &RI.

Marine Biodiversity

Pondicherry territories are situated on the coast and quite a sizeable population depends on the exploitation of marine resources. Over exploitation of fisheries has major impact on marine biodiversity leading to impoverishment of stocks on target species. This can have a knock-on effects to other aspects of the marine eco system, including its productive capacities. However fisheries being an open access and mobile resource, which makes the definition and enforcement of limits problematic.

Introduction of concrete conservation measures and regulation and enlisting of endangered fishes and prawns.

Afforestation programmes of mangroves as conservation of natural nurseries for prawns and fishes.

Implementing welfare schemes for fisher folk during fishing holidays.

Tapping of indigenous knowledge.

Implementing strict measures and close monitoring of the factories near coastal area.

Providing biological sea wall along the coast

Proper monitoring of fishing holidays.

Recommendations

The rights of native fishermen-to traditional livelihood and preferential access to marine resources are increasingly affected by competitive interests such as industrial fisheries and to protect their interest do not encourage large scale fishing.

Local fishermen communities should be empowered in the decision-making process through the implementation of community-based approaches to fisheries management.

Resource management decisions must be ecologically-based, conservation-driven and risk-averse.

Fish food units should be established where women folk can be engaged and an avenue for income generation be opened.

Women self-help groups should be formed and train them in rearing ornamental fish production which is in demand now.

Specific management objectives and quantifiable measures of progress towards sustainable fisheries should be developed.

The fisher folk must be effectively integrated into the eco system through cultural and community awareness, education, ethical responsibility and adequate legislation and enforcement.

A process should be developed for identifying prioritizing habitat protection and restoration opportunities and for implementing the associated activities.

The long term value of habitat must be recognized and fully appreciated by every one involved in resource management and use.

Government agencies and communities based organizations must be funded and staffed.

NGOs and social groups should also be involved in addressing the concerns and issues relating to the fisher folk.

To maintain the stocks, biodiversity and environment incentives should be given for responsible fishing. Awareness should be created among the fishing community about marine biodiversity and the damages done by them unknowingly, through film shows, issuing leaflets in local languages. The fishermen co operatives should be vested with the responsibility of protecting fisheries resources they exploit. They should be made aware of the biological and the environmental basis for sustainability of fish stocks, by constant interactions with scientific communities responsible for the same, which will be beneficial to the fishermen, making implementation of the management options smooth and effective. The Regulation and conservation rules should be farmed in consultation with the Law department and the Department of Science, Technology and Environment and these rules and regulations should be let known to the fishing community.

A project should be undertaken to enlist the endangered fishes and prawns of Pondicherry region, under the supervision of a Joint Director, Fisheries Department. The work could be done in collaboration with the Pondicherry University and the local Degree and Postgraduate colleges.

The traditional fishermen have an indigenous knowledge about the seasonal availability of fish and fish catch, quality of fish, hydro geographic relations of fishing, breeding seasons, (when not to catch), which will be very valuable in the biodiversity conservation. The generations knowing the traditional fishing methods which are disappearing need to be tapped to train the younger generation. The fishermen who has the knowledge of traditional fishing should be identified by the department and they should be made use of to train the younger generation. The fishermen who train the younger generation can be suitably rewarded.

Necessary steps should be taken in consultation with the Department of Forests and Wildlife to develop mangrove plantation along the estuaries which will serve as a natural breeding ground for fishes and prawns. In addition the plantation will serve as a wind breaker to the coastal villages during cyclone.

Awareness should be created about the importance of mangrove plantation among the villages in the form of slide shows and issuing leaflets. The plantation should be maintained by the community by giving incentives who takes care of the up keep of the plantation. In addition to the support measure taken by the Department of fisheries such as 25 Kgs of rice during ban period, the Department should involve womenfolk in training them to do handicraft and should try to make them financially secure through providing training in pickling of fish and other activities.

During the public hearing conducted in Thirumalairayan Pattinam, Karaikal region, it is understood that the coastal pollution in that region causes human and animal health hazards and reduction in fish harvest. The Department of Fisheries and The Department of Science, Technology and Environment should take necessary steps to prevent further damage.

To prevent soil erosion, biological sea wall should be established along the coast in consultation with Agriculture department by identifying the spots where there is more erosion of soil.

The Coast Guards may be requested to monitor the coastal area against illegal fishing in order to enforce the fishing holidays effectively. Proper mobilization equipments may be provided for monitoring.

Inland Aquaculture should be done in water spreads in villages, where water is not used for fish rearing or any other productive activity. The advantage of utilizing the unutilized resource like a pond or tank in a village helps in resource base being upgraded and profitability used to provide income and employment for the resource poor. Also fish ponds in villages act as a drainage center for the run off water for that village.

Implementing Agencies : Department of Fisheries, Department of Agriculture, Department of Forests & Wildlife and Department of Tourism and Department of Port. Budgetary provision should be made for incentives, awareness programmes, mangrove plantation and handicraft training.

Education

There has been no concept of biodiversity related issues in the education system being followed in Pondicherry. Education related to conservation of biodiversity has to be started in school training of primary, secondary and tertiary level teachers on nature conservation and environmental protection, has to be undertaken along with task of preparation of suitable teaching materials. It should lead to a wide range of activities making use of all possible means of communication in creating awareness among children about biodiversity.

Inclusion of biodiversity related education in primary / high school and higher secondary school curriculum in a graded manner and non formal manner.

Integrating biodiversity themes into environmental education units for all pre-and in-service teacher training courses and B.Ed. courses. Provide training in the use of practical, field-oriented methods of teaching biodiversity.

The introduction of biodiversity related issues should be started from the first standard in a very friendly manner. The issues should be brought with the help of stories, films, field visits etc,. The students as per their age groups should be expected do some activities in their neighbourhoods. The problems should be analyzed in an environment conducive to the students. It may follow a non-formal method of studies.

In the colleges and Universities environment protection, environment and biodiversity related issues should form an integral component of all the syllabi as per the recommendation of the University Grants Commission. However, the involvement of the students in community development and conservation of biodiversity should be ensured through NSS and NCC programmes also. Street plays, Tree Plantation Programmes, Clean and Green Pondicherry Programmes could be important component of these programmes.

Teachers at schools are not familiar with the concept of biodiversity. The priority is to develop appropriate short courses and introduce these courses to teachers particularly in primary and secondary schools. Biodiversity related topics should be integrated in the education units for all pre teachers training courses and B.Ed. courses, providing field and practical training. Awareness and acknowledgement of biodiversity conservation by teachers is the first important step for biodiversity. The Department of Education should give more emphasis for biodiversity component while conducting exhibitions and festivals in schools and colleges annually. All the schools should be instructed to conduct biodiversity related activities and various awards should be instituted at primary / secondary/ college levels by conducting elocution, essays, quiz competitions. Some important days like " National Science Day" "World Environment Day," "World Earth Day," " Human Right's Day" etc., should be used to address various environment and biodiversity related issues. The Nature clubs in schools should be encouraged to grow rare medicinal plants, indigenous species in their school premises garden as a part of biodiversity conservation the whole focus should be towards "Greening" of the education. During preparation of programmes of art and cultural performance, experts of biodiversity conservation can be invited to provide technical guidance and consultation to improve the quality of the programmes.

Special posters and placards on biodiversity conservation should be installed on specific sites including main road crossings, parks, botanical gardens, museums, and beach. These posters and placards can be used to introduce the basic knowledge on protected species, habitats, main laws and regulations of biodiversity conservation and others. These special sites can be used as in-situ classrooms for students in primary and secondary schools to present knowledge of biodiversity conservation to them.

Professional training to managerial and non-professional staff : Decision-making and ordinary non-professional staff should be given training on the biodiversity conservation which causes a gap between the economic development and the biodiversity conservation. In many cases the public, mass media and education staff have very limited knowledge of biodiversity conservation. All managerial staff in the governmental agencies and institutions lack specific knowledge on biodiversity conservation. Sensitization of social workers,. politicians and legal experts on biodiversity related issues should also be undertaken.

A biodiversity conservation training center for decision-making managerial staff at the ministerial and department levels should be established under the Department of Science,

Technology and Environment. All relevant staff could be trained at least once within three year time period.

Funds should be generated by preparing a budget for the ensuing financial year in consultation with the Department of Planning and Research and Department of Finance for implementing various biodiversity related activities.

For some economic issues and economic planning, experts of biodiversity conservation, economists from concerned departments / university can be invited to join such events to make technical presentations or consultations so that the biodiversity conservation contents can be incorporated and included in those programs and projects.

Implementing Agencies : Department of Education & Department of Science, Technology & Environment, Government of Pondicherry.

Animal Husbandry

Livestock diversity has faced a serious threat in Pondicherry. The greatest factor in the loss of domesticated animal diversity has been deliberate cross-breeding with exotic breeds, carried out extensively by the government in order to increase the yields of milk and other animal products. Semen banks have generally stored the semen of exotics. The policy thrust towards maximizing financial profits has severely threatened species and breeds which were bred for a variety of domestic uses. The breeds like Sindhi, Tharparkar, Hallikar, Gir etc. ideally suited for the climatic conditions and fulfilling local requirements of draught, dung and milk, has been pushed out by ' high' milk producing cross-breeds and exotics.

The Aseel poultry. bred for meat, cock-fighting, some egg production, and easily managed by poor rural women, has been displaced by exotics which are better suited for large-scale commercial farming. While all kinds of livestock are affected by cross-breeding, worst off is poultry; exotics now make up 80% of the total poultry population, with disastrous effects on indigenous breeds. Export oriented poultry production is likely to intensify the loss. The Union Territory of Pondicherry does not have a recognized breed of its own. Before the introduction of cross breeding programme, the desi breeds of neighboring Tamil Nadu State were said to be prevalent. As cross breeding programme is popular with the public, the Government of Pondicherry introduced exotic semen in the breeding programme in order to increase milk production.

Threat : Cross breeding the local breeds with exotic semen with a view to increase milk production.

Strategies : Protection and propagation of desi breeds of domesticated animals as these are more tolerant to local stresses and for their potential use in future breeding programmes.

Establish Breed Conservation Centers and publishing of breed wise information and its importance for awareness of local public and farmers.

Action plan : Public awareness should be created about the importance of the local breeds by publishing leaflets, booklets and film shows. During annual poultry shows, more emphasis should be given to encourage the breeder of the local and indigenous breeds and cash awards should be instituted for the best local breeds. The Department should formulate a scheme to bring back the suitable Indian breeds which can withstand stress and disease attacks. The scheme will help the poor farmers to make use of the male calves thus produced, for ploughing purpose during agricultural seasons instead of farmers engaging tractors etc., at a higher rent and sending these male calves to slaughter. Frozen Semen of the local breeds should be readily made available for the farmers in the Breed Conservation Centers.

To conserve and reintroduce the local and indigenous breeds of poultry, the chicks of local breeds should be distributed to the rural farmers under backyard poultry schemes which will help the rural poor to generate income at a minimal cost of maintenance, and will conserve the biodiversity.

Implementing agency : Department of Animal Husbandry.

Forests & Wildlife

Union Territory of Pondicherry is devoid of natural forest. Under the social forestry programmes since 1986 forest plantation have been raised along road sides, school campus, parks and in poramboke lands.

As there is no forest an intensive project should be taken up for plantation activity in urban and rural areas. Development of mangroves in Pondicherry, Karaikal and Yanam regions. Conservation of the sacred groves in Pondicherry and Mahe region.

Mangroves-Pondicherry Region : Due to population growth and overexploitation, the mangrove vegetation in Pondicherry is rapidly degraded and depleted in the last two decades. There are four regions of Pondicherry Coast with mangrove vegetation which is overexploited, degraded and depleted. These areas are: 1. Thengaithittu. 2. Veerampattinam 3. Ariankuppam and 4. Murthikuppam.

To restore the mangroves sustainably in Pondicherry regions the following actions are required.

Vegetation profile Establishment : The mangrove vegetation profile of Pondicherry region is poorly understood. The taxonomy of mangrove species of different locations should be established. A mangrove vegetation map should be prepared which will provide the mangrove vegetation scenario of Pondicherry.

Avicennia is the dominant genus in all locations : On-site research and demonstration should be undertaken to grow the other economically important mangrove species including, Bruguiera, Excoecaria, Rhizophora and Acanthus. In addition, the growth patterns of other mangroves should also be studied in different locations. Plantation should be undertaken to accelerate the reclamation of the soils near the Veerampattinam village.

Phenological Studies of Mangroves : The phenology and morphometric data of different mangrove species should be carried out to understand the growth potential of different species at different sites. These include the net growth, internodes, leaf growth and leaf area which would provide the basic information on the plant growth performance. The above studies will provide all the information for the growth and establishment of different mangrove species at different locations around the coastal belt of Pondicherry. This data will be used to formulate a specific growth model for the mangrove species. Thus, scientifically managed growth and silvicultural plots will be established to provide growth and biological data.

Interaction with Village groups : After establishing the above said data and basic growth characteristics of various mangrove species, interaction with village groups has to be undertaken to enthuse them with all information related to the importance of conservation of mangrove species.

Seminars/workshops and public hearings : Seminars / workshops and public hearings should be conducted for various users of the mangrove environment including farmers, fishermen and shrimp farmers as the target groups with a strong message that they will be involved in protecting their own resources.

As this programme is primarily aimed at natural resources utilization and management strategies of Pondicherry Coast, the target groups will be mostly rural farmers, fishermen, shrimp farmers. They should be enthused in conservation of mangroves. Participation of rural communities in mangrove-based plantation activities should be given priority. It is also to be intended to educate local public to encourage them to a degree of self management amongst the various users of the mangrove environment, so that they are involved in protecting their own resources. Mangrove management plans, no matter how well articulated, cannot succeed without regard to the requirements and aspirations of indigenous and local communities. We should thus plan to educate the local populations through extension and secure their support and commitment in establish in the depleted mangrove vegetation.

There are several environmental concerns regarding the use and management of mangrove ecosystems. The concerns will vary from region to region or country to country in terms of severity and areas of impacts. However the concerns are severe at the local level where remedial measures are very much necessary. The most common concern is deforestation of mangrove vegetation and consequent depletion of fisheries and other mangrove dependent living resources. Also serious degradation or destruction of critical habitats associated with loss of biodiversity and / or genetic resources. Thus there is immediate need to restore the depleted mangrove vegetation to protect the environment and ecosystem.

Overexploitation of mangrove forests occurs mainly due to primary lack of awareness of several problems rather than to a lack of economic alternatives. In addition to establish the site-specific mangrove species, the main objective of this program should also include to win public acceptance and support for this type of research programs. The final output should be to restore the degraded mangrove vegetation and re-establish the glory of the mangrove forest and finally to çonserve the biodiversity of mangroves in Pondicherry region.

Restoration of mangroves in Yanam and Karaikal Regions : To develop further the mangrove vegetation in Karaikal and Yanam regions plantation of mangroves should be taken extensively along the coast. Strict law should be enforced to protect the existing mangroves as there is human interference.

Implementing agency : Department of Forests & Wildlife and School of Life Sciences, Pondicherry University, will coordinate and execute the action plan, and The Department of Forests & Wildlife would fund the project.

Sacred Groves

For successful conservation of sacred groves the following points of action should be implemented.

- Cultural traditions (belief & faith) to be encouraged
- Create awareness on ecology and environment of the vegetation
- Develop buffer zone to avoid pressure from groves from collection of fuel, food and fodder
- Develop buffer zone to avoid pressure from groves from collection of fuel, food and fodder
- Setting up an interpretation center regarding cultural connections relating to conservation of biodiversity

- Establishing Village Forest Council (VFC) for planting, maintenance and protection
- Conventional energy systems like solar, wind etc. should be encouraged

The following sacred groves with fair indigenous vegetation in the Union Territory of Pondicherry should be given priority in conservation.They are, Sacred groves in Pooranankuppam, Keezhkumaramangalam, Manapet, Sedarapet, Karasur, Mangalam, Moorthikuppam (Thennampakkam).

Sacred Groves in Mahe Region : Ayyappan Kavu in Pandakkal and Chembra hills, Mahe have dense vegetation which has to be enumerated and preserved. Such patches should be declared as protected areas in consultation with the committees of Ayyappan Kavu to ensure a legal back bone for conservation effort.

Implementing agency : Department of Forests and Wildlife and village communities.

Swadeshi Cotton Mill campus : Swadeshi Cotton Mill campus is the only patch of forest available in Pondicherry. During the process of preparation of State action plan a patch of Swadeshi mill land was given for the construction of court building. A list of the flora existing in Swadeshi mill compound area was prepared and is given in appendix. Since no evaluation of loss of biodiversity after clearing a patch of land was done. It is suggested that such an evaluation should be made and steps to restore lost species, if any should be made.

Pollution

Solid Waste Management

Awareness should be created among the public, specially to the house-wives, school children about the proper disposal of Domestic Water / solid wastes etc. Awareness campaigns should be periodically conducted in the developing colonies and in existing colonies.

The existing system of door-to-door collection of solid wastes in many areas should be introduced in all the residential and business areas. However colour coding of bags for separating biodegradable and non-biodegradable wastes should be introduced at the earliest. This will help the authorities in separating the wastes for composting. Plastic materials should be banned to be used for land fills.

Industries should be monitored closely for proper implementation of the waste disposals and having effluent treatment plant installed. Erring industries should be penalized. The Biomedical Wastes from the hospitals, laboratories and slaughter houses should be handled separately and professionally trained personnel should be used for this job. Incinerator facility has to be upgraded.

The commune Municipalities should identify places for disposable of organic wastes which can be used for vermin-composting. Local working groups, NGOs must be encouraged to help handling waste.

A proper drainage system should be developed to dispose the domestic water waste and before its disposal into sea it should be treated and the treated water can be made use of for growing fodder.

The staff on duty and other personnel handling garbage should be given proper orientation by the resource persons qualified in the waste management and health aspects. The issue of health of waste pickers should be seriously addressed and they should not exposed to the spillage of garbage during transportation and also while handling it. Protective clothing and periodic health check up is to be ensured.

Implementing Agencies : Pondicherry Pollution Control Committee, Department of Science, Technology & Environment, Department of Local Administration/ Municipalities, Department of Health and Family welfare.

Bio Medical Waste Management

Awareness through various media should be created amongst patients, attendants medical personnel and people in general regarding hazards of bio-medical waste management. Proper waste management from hospitals, nursing homes and laboratories must be ensured and they should asked to get their waste management plans approved as per their guidelines to be issued to them. National standardization of color and coding should be followed for handling biomedical waste and this will be helpful to eliminate hazard from waste and its adverse effects on health.Training programs for all sanitation personnel in hospital waste management should be made mandatory.

An integrated approach with sharing treatment and disposal technology amongst a group of hospitals and medical units should be introduced. This will be cost effective and can cover more number of medical units.

A hospital and laboratory waste management task force should be formed. This should consist of Health Secretary, Director of health Department, Director, JIPMER, scientists from Pondicherry University,medical experts, Non Governmental Organizations, private sector, representatives of community and municipalities.

Incentives should be provided to medical and research units for installing incinerators and equipments for treatment and disposal of hospital waste.

Implementing Agencies : Pondicherry Pollution Control Committee, Department of Science, Technology & Environment, Department of Local Administration/ Municipalities and Concerned Hospitals, Laboratories, Research and Educational Institutes.

Slaughter House Waste

More slaughterhouses have to be provided in various areas and make shift slaughtering of animals should be completely banned.

Modern or mechanized slaughtering of animals should be done to maintain a more hygienic environment.

Unauthorized slaughtering of animals should be totally prohibited in the rural and urban areas. Strict enforcement of should be carried out and erring persons should be penalized.

Fish market waste should not be mixed with municipal solid waste. It must be collected and buried separately after taking the necessary treatment measures.. Fish market waste can be used to manufacture poultry feed. The slaughter house waste should be collected and dumped in separate pits meant for the purpose.

Non-biodegradable Waste

The plastics and other non-biodegradable waste should be collected and handled separately and efforts must be done to get them recycled with the help of private contractor or putting a recycling plant in Pondicherry.

The use of plastics should be minimized in the Government organizations like Amudhasurabhi, Vanavil, Le Café etc. Gauge measuring machine should be installed and anybody found using plastics of less than 20 microns should be punished.

Through the help of various awareness campaigns including the involvement of Rotary and Lions club, the establishments not using the plastic bags should be given a special status. Consumers should be encouraged to bring cloth bags for carrying their groceries and other purchases. Shopkeepers should be encouraged to allow cash discounts / small gifts to the customers bringing their own shopping bags. The prices should be fixed higher with the usage of plastic cups.

Awareness campaigns should be initiated aggressively to educate people with the various aspects of plastic pollution. Warnings may also be mandatory 0n all plastic materials including toys.

No fresh licenses to be granted by the Government of Pondicherry to start units manufacturing plastic bags, packing materials, PET bottles, toys etc., Instead encouragement must be provided to the industries set up for recycling waste without polluting the environment. Production of paper cups and earthen pots may be encouraged.

Heavy penalty should be fixed if the garbage of hotels, Kalyan mandapams (Wedding Halls) contain plastic wastes. Penalty should be fixed for littering public places with plastic bottles, bags etc.,

Phased removal of plastic use and providing of alternate employment opportunity should be provided to the employees working in plastic industries.

Tourism

The Government should focus 'eco-tourism' policy to encourage eco-friendly practices that focus on harmony with nature and spirituality and to develop the conscious process at all levels. Pondicherry has already made its name in spirituality and Yoga. For the several years Pondicherry has been hosting International Yoga Festival in the first week of January. The Government should plan opening of Yoga centers and spirituality centers so that sustainable tourism could be addressed and involvement of local persons could be there without degrading biodiversity.

However, the Government should aim to encourage low volume but high value tourism in Pondicherry, which could be directed to eco-friendly tourism. The mangrove vegetation in Yanam is so beautiful and should not be disturbed and the damaged patches could be restored. Usage of environmental friendly constructions would not help conserving biodiversity. Any activity damaging the natural eco-system in islands near Yanam should not be allowed.

In Pondicherry Ousteri lake could be developed as an 'eco-friendly' tourist spot and local communities should be involved in the whole process. It is necessary that the water inlets to the lake be opened so that the lake could get enough water. Fruit bearing and shade trees could be planted near the lake so that migratory birds could start visiting again during winter.

In places like Botanical Garden a fee should be collected to restrict the human activity which damages biodiversity. In such cases only the interested public will visit and the fee could be used to maintain the garden. No food should be allowed inside the garden area.

Apart from the revenue generated at tourist places, they could be used for biodiversity conservation in those places and assure sustainability.

Town and Country Planning

Land ceiling Act has been a key factor for the conversion of agriculture lands, wetlands in housing and industrial sites. Now after repealing of the Act, Town and Country Planning

Department must strictly enforce laws to prevent conversion of agriculture land into urban areas. The existing wet lands and tanks should not be filled in the name of urban development as this has been a key cause of biodiversity destruction. Accordingly the Government should formulate a strict law to protect the above said lands. Law enforcement and legal system needs to be sensitized to check such acts.

Pondicherry as a heritage city, has a strong French culture. There are many buildings which were constructed at the time of French rule. Pondicherry has a rare mix of French and Tamil history that makes the town unique. The "White" and "Tamil" part of the town are still in splendid shape in many areas. The old French buildings in the white town there are many buildings such as Marie, Raj Bhavan, Sacred Heart Church, Cathedral, Sri. Aurobindo Ashram and Museum which have unique architectural significance. In addition they provide place for protection of biodiversity in the form of gardens and kitchen gardens. It has been noticed that once such buildings are demolished the new structures which come up at these places are only concrete structures covering the whole ground areas. The high land prices in the white town areas have seen the demolition of several protectable buildings. This has caused extensive damage to biodiversity in Pondicherry.

The Government should immediately declare all the buildings constructed during French regime as heritage buildings and stringent measures have to be adopted for their preservation and maintenance. Use of such buildings for biodiversity conservation must be ensured. Any additions, alterations, renovations in these buildings must follow strict measures so that the architectural and biodiversity heritage of Pondicherry could be saved.

Rain Water Harvesting

During the past years due to increased urbanization, industrialization and over exploitation the ground water table has been decreasing at an alarming rate. Though there are regulations of rain water harvesting for houses, they are hardly been followed, causing a serious threat to this precious natural resource. Non availability of ground water in several parts of

Karaikal and Pondicherry has been a cause of depletion of biodiversity in addition to other problems faced by the residents.

Pondicherry Union Territory experiences moderate to heavy rain fall and most of the precious rain water goes to the sea. It is imperative that the rain water harvesting policy is redefined and strictly enforced.

Rain water that falls on terrace of the houses and buildings could be diverted to the existing open wells / borewells.

Percolation pits should be provided in all the small houses. All the house owners must be given a dead line to comply with it. Heavy fines should be levied for the persons who do not provide for rain water harvesting within two years from the date of notification.

In the big houses and apartment buildings recharge trenches have to be provided for. In the large buildings and industries Recharge wells must be dug so as to harvest rain water. To start with all the Government buildings, educational institutions should become model examples for rain water harvesting. Rain water harvesting systems should also be installed at public places so as to popularize the concept.

Incentives in the form of tax deduction, house tax, water tax etc., could be provided to induce and reward the persons. Banking loans could be arranged for fixing rain water harvesting system.

Implementing agencies : Department of Town & country planning, Law Department and Department of Agriculture

Reduction of Traffic Density

Staggering of office / school and factories timings would significantly help to reduce the otherwise peak hour pressure on the roads. Similarly there is an immediate need to strengthen the existing public transport system so that this would help public to reduce using their own vehicles.

Biodiversity Board

To conserve Biodiversity and to control its related activities a State Biodiversity Board has to be established in the Union Territory of Pondicherry. The Board will consist of representatives of various Government Departments, Academics, Scientists and NGOs.

The board could be constituted away from the government set up and the Chairman should be an eminent person having adequate knowledge and experience in the conservation and sustainable use of biological diversity and in matters relating to equitable sharing of benefits.

The State Government may nominate ex-officio members from the Department of Agriculture, Department of Fisheries, Department of Science, Technology & Environment, Department of Tourism, Department of Education, Department of Forests & Wildlife and Pondicherry University.

Members

In addition the Government should nominate coordinators of Biodiversity Action Plan and also NGOs and women representatives.

The functions of the State Biodiversity Board could be as follows.

To advice the State Government, subject to any guidelines issued by the Central Government, on matters relating to the conservation of biodiversity, sustainable use of its components and equitable sharing of the benefits arising out of the utilization of biological resources.

To regulate by granting of approvals or otherwise requests for commercial utilization or bio-survey and bio-utilization of any biological resources by Indians.

To coordinate the biodiversity conservation related projects submitted by various organizations.

To perform such other functions as may be necessary to carry out the provisions of this Act or as may be prescribed by the State Government.

Finance, Accounts and Audit of State Biodiversity Authority

The State Government may, after due appropriation made by the State Legislature by law, may pay to the State Biodiversity Board by way of grants or loans such sums of money as the State Government may think fit for being utilized for the purposes of Biodiversity Act.

1. There shall be constituted a fund to be called the State Biodiversity Fund and there shall be credited thereto
   a. any grants and loans made to the State Biodiversity Board under section 31
   b. any grants or loans made by the National Biodiversity Authority
   c. all sums received by the State Biodiversity Board from such other sources as may be decided upon by the State Government
2. The State Biodiversity Fund shall be applied for
   a. the management and conservation of heritage sites
   b. compensating or rehabilitating any section of the people economically affected by restriction imposed under section 37
   c. conservation and promotion of biological resources
   d. socio-economic development of areas from where such biological resources or knowledge associated thereto has been accessed subject to any approval granted under section 24, in consultation with the local bodies concerned
   e. meeting the expenses incurred for purposes authorized by this Act

The State Biodiversity Board shall prepare, in such form and at such time each financial year as may be prescribed, its annual report, giving a full account of its activities during the previous financial year, and submit a copy thereof to the State Government.

The accounts of the State Biodiversity Board shall be maintained and audited in such manner as may, in consultation with the Accountant-General of the State, be prescribed and the National Biodiversity Authority shall furnish, to the State Government, before such date as may be prescribed, its audited copy of accounts together with auditor's report there on. The

State Government shall place the annual report and auditor's report as soon as they are received, before the House of State Legislature.

Implementing agency : Department of Science, Technology & Environment.

Pondicherry University

Establishment of Medicinal Plant Park

It is now known that 1500 drugs are of plant origin that have been used to cure human ailments. However, the pharmacological properties are known only on a few number of medicinal plants. In addition exploitation and ruthless collection of several medicinal plants from natural resources resulted a rapid depletion of several medicinal plants.

Availability of raw material obtained from the medicinal plants is most essential to provide quality medicines. Therefore, we should look for increasing the availability of medicinal plants.

Although several medicinal plants have been reported from Pondicherry region the following medicinal plants are presumed to be threatened in the wild.

1. Alpinigia galanga,
2. Acorus calamus,
3. Gloriosa superba,
4. Ormocarpum sennoides,
5. Saraca asoca,
6. Gymnema sylvestre,
7. Amorphallus sylvaticus

Hence, collaborative programmes between the Department of Agriculture and Pondicherry University should be established to encourage the cultivation of medicinal plants should be strengthened.

Also coordination between Pondicherry University and state Forest Departments should be encouraged for taking up programmes of growing medicinal plants. These activities are to be scrutinized about the plants which are in short supply and which need to be propagated. Short term training programmes on collection and harvesting of medicinal plants should be introduced.

Thus there is a need for proper coordination among the Department of Agriculture, Department of Forests and Wildlife, Government of Pondicherry and Pondicherry University for overall development of endangered medicinal plants.

Standing committees with representatives drawn from the concerned departments and coordinators of NBSAP, Pondicherry University are recommended.

8

Freshwater Resources and their Conjunctive Use

The world's fresh water resources are unequally distributed both in time and in space. Until recently water resource management focused on reallocating water to when and where it was required, a supply-side or fragmented approach. Nowadays there are signs that water resource availability is dwindling-due to both population growth and increased per capita water use-and ecosystems are being damaged. To face this challenge a new holistic approach is needed. This approach includes the integrated or conjunctive use of surface and groundwater resources and takes account of social, economic and environmental factors. Moreover, it recognizes the importance of water quality issues. In this context, the paper examines the main aspects and problems concerned with the planning, design, construction and management of conjunctive use of surface and subsurface water resources, along with its environmental impacts and constraints to sustainable development. The importance and role of research thrust, technology transfer, institutional strengthening, effective partnerships between governments and stakeholders, and sound financial frameworks are also examined. Finally, the challenges and benchmarks for future actions that the scientific community and planners have to face and deal with, are briefly outlined.

INTRODUCTION

Water resource management should preserve or enhance the environment's buffering capacity to withstand unexpected stress or negative long-term trends. As the environment's carrying capacity is put under increasing pressure, due to the growing needs of the population and improper use of its resources, environmental vulnerability increases too. In this context, mismanagement of water resources, paying only lip service to the environment, has led to water scarcity and water pollution which threaten security and the quality of human life. Giving proper regard to this unsustainable trend, the Second World Water Forum acknowledged the pivotal role that integrated water resource management plays in the process of sustainable development. The term "integrated" embraces the planning and management of water resources, both conventional and non-conventional, and of land. It takes account of social, economic and environmental factors and comprehends surface water, groundwater and the ecosystems through which they flow. Moreover, it recognizes the importance of water quality issues.

Integrated water resource management depends on co-operation and partnerships at all levels, from individual to governmental and non-governmental, national and international organizations sharing a common political, scientific and ethical commitment to the need for water security and for optimizing water resources use and planning. To achieve this goal, there is a need for coherent national, regional or interregional policies to overcome fragmentation, and for transparent and accountable institutions at all levels. To this end, targets should be established and suitable strategies should be devised to meet the challenges inherent in the sustainable use and development of water resources. These resources should be managed at both the river basin and aquifer levels.

The management framework should envisage a high level of autonomy for the body responsible for water use and planning, allow for stakeholder participation in decision-making and generate and disseminate information. Where appropriate, specific river basin, catchment and groundwater authorities should be set up, and their capacities enhanced. Where water is shared, actions should be taken to build confidence among riparian states, enabling them to accept some form of restricted sovereignty regarding their common resource, based on both equitable utilization and regional cooperation. Besides institutional strengthening, sound and fair financial management, based on the "user pays" principle is needed to improve the efficiency of services, provide additional resources for investment, encourage demand management, and promote pollution control and prevention.

Research must be directed towards solving water use and planning problems, gaining a better understanding of the hydrodynamic and hydrochemical processes involved and enhancing water productivity. Action research should cover field and laboratory evaluation, assessment and monitoring, development and implementation of suitable water management strategies. This process requires enhanced basic and applied research and a large variety of tools ranging from field techniques to advanced technology for water control and regulation such as models, Remote Sensing, Geographic Information Systems, Decision Support Systems and spatial analysis procedures. All these tools have to be considered under a broad and integrated approach for addressing the use, planning, conservation and protection of both surface and subsurface water resources, that takes proper account of the environmental impacts and socio-economic effects of development.

THE CONCEPT OF CONJUNCTIVE USE

As broadly outlined above, a critical problem that mankind has to face and cope with is how to manage the intensifying competition for water among the expanding urban centres, the agricultural sector and instream water uses dictated by environmental concerns. Confronted with the prospect of heightened competition for available water and the increased difficulties in constructing new large-scale water plants, water planners must depend more and more on better management of existing projects through basin-wide strategies that include integrated utilization of surface and groundwater. Todd (1959) defined this process as conjunctive use. Lettenmaier and Burges (1982) distinguished conjunctive use, which deals with short-term use, from the long-term discharging and recharging process known as cycle storage. Until the late fifties, planning for management and development of surface and groundwater were dealt with separately, as if they were unrelated systems. Although the adverse effects have long been evident, it is only in recent years that conjunctive use is being considered as an important water management practice.

In general terms, conjunctive use implies the planned and coordinated management of surface and groundwater, so as to maximize the efficient use of total water resources. Because of the interrelationship existing between surface and subsurface water, it is possible to store during critical periods the surplus of one to tide over the deficit of the other. Thus groundwater may be used to supplement surface water supplies, to cope with peak demands for municipal and irrigation purposes, or to meet deficits in years of low rainfall. On the other hand, surplus surface water may be used in overdraft areas to increase the groundwater storage by artificial recharge. Moreover, surface water, groundwater or both, depending on the surplus available, can be moved from water-plentiful to water-deficit areas through canals and other distribution systems. On the whole the integrated system, correctly managed, will yield more water at more economic rates than separately managed surface and groundwater systems.

Water Storage

In conjunctive use, the two most important issues that planners have to face concern the storage of surplus water and the optimal allocation of water withdrawals.

With regard to the first problem, a question that needs to be answered is where to store water and which reservoirs to develop: surface or subsurface?

Subsurface Water Storage

The advantages of subsurface over surface reservoirs are:

- surface reservoirs are lost forever once they are silted up, while underground storage capacities remain practically unaffected by development;

- yields from groundwater storage, less affected by evaporation and leakage, are more dependable than yields from surface reservoirs;
- groundwater is less prone to pollution than surface water, and if polluted, pollutants can be diluted during underground movement;
- subsurface storage is achievable without loss of water-spread areas suitable for cultivation or other beneficial land uses;
- groundwater can be put to use where and when it is required, with less risk of seepage or evaporation losses during storage and transmission;
- there is less ecological hazard compared to surface storage projects;
- groundwater storage is less liable to deterioration than surface storage;
- the cost of storing groundwater is less than that of surface storage.

In spite of the many advantages mentioned above, there are some constraints that hinder groundwater storage, such as:

- wells interfere adversely when large supplies are required;
- groundwater storage withdrawal is a highly energy intensive process, while surface water is often available by gravity flow;
- surface reservoirs are more suitable for multiple uses, including energy production and recreation;
- mineralization is generally lower in surface water storage.

The current trend in aquifer management focuses on determining the maximum and minimum water levels, in order to regulate storage capacity. As a matter of fact, uncontrolled overexploitation causing progressive drawdown below the minimum permissible piezometric levels, will lead to increased pumping costs, land subsidence, infiltration of poor quality water, drying up of springs and shallow wells, decreased river flows. Moreover, in coastal aquifers the prolonged reduction of freshwater flow towards the sea reduces the equilibrium gradient, inducing saltwater intrusion and the inland movement of the freshwater-saltwater interface. Combining so many aspects requires methods of analysis that systematically integrate them in such a way that within the planning process alternative solutions can be defined, tested and chosen.

Normally artificial groundwater recharge is accomplished by means of infiltration basins or injection wells. Other techniques for augmenting subsurface supplies include vegetation management, runoff inducement and increasing seepage from streams by widening the wetted perimeter of channel sections or lowering the groundwater table in the flood plain.

Water quality aspects play a major role in this process. They mainly concern the quality of recharge water and its effects on groundwater quality. One striking example is the "Water Factory 21" plant in Orange County (California), where wastewater undergoes an advanced treatment process before being injected into deep wells to create a barrier against seawater intrusion (Cline, 1983).

Generally speaking, numerous issues need to be addressed before suitable recharge systems can be chosen, designed and managed for optimum environmental and economic performance. One problem is proper site selection, which requires field surveys and infiltration/soil hydraulic conductivity measurements to predict seepage rates. More research is also needed on optimum management of storage systems, including flooding and drying schedules for infiltration basins, as well as pre-treatment (sediment removal) of water.

To address the above-mentioned groundwater management problems, the following steps or phases should be considered and carried out:

- general groundwater surveys and identification of the sites that require in-depth studies; these studies provide estimates of water quality and quantity, corroborated up by reliable data;

- geohydrological investigations aimed at determining more accurately groundwater availability and quality in terms of time and space, using mathematical analysis in order to establish aquifer conditions and behavior;
- integration of the physical characteristics and conditions previously collected and analyzed, with economic and social parameters to formulate suitable strategies and policies for subsurface water use, planning and management.

Surface Water Storage

For surface reservoir management the critical elements to be considered are minimum pool elevation and storage losses due to sedimentation. Generally, minimum pool elevation is not defined solely by hydraulic limitations of the outlet or diversion works; more severe constraints may be imposed by recreational interests, habitat values in the reservoirs or by the adverse water-quality effects if the pool is drawn too low. Loss of storage due to silting is normally significant only if based on projections of 50 or 100 years, so regular sediment surveys (at least once every 10 years) are important aspects of the process.

To account for these factors, generally, a two-step design process is adopted (McMahon, 1992). In the first step, a number of potential reservoir sites are examined, not only for construction requirements, but also in terms of hydrologic patterns in order to establish capacity-yield relationships. This procedure leads to the "preliminary design" framework. In this phase simplifying assumptions are normally made: reservoir releases are assumed constant, evaporation is ignored in temperate and humid regions, seasonal flows may not be taken into account, and so on.

In the second step, leading to the "final design", procedures must account for all factors affecting the project design, including fluctuations of inflows and release by season, release restriction during periods of low storage, evaporation losses, minimum pool requirements and supply failure probability. Uncertainty is a major element of concern in the design process. It not only affects flow records, where temporal and spatial variability is significant, but also the generation of demand forecasts.

Allocation of Water Releases

Linked to storage is the optimal allocation of water releases. Heightened competition for withdrawals, increasing in-stream flow regulations, compelling groundwater quality issues, along with environmental concerns, lead to the formulation of permitting programs and the establishment of regulatory agencies aimed at coordinating and controlling water resource allocations. The first task in this process is to determine and explicitly formulate the overall goal of the permitting systems and to establish permitting rules that reflect those objectives, such as: maintenance of in-stream flows, economic development, water rights and protection of surface and groundwater bodies.

A great variety of mathematical techniques has been proposed to solve problems of optimal allocation of water withdrawals. Eheart and Lyon (1983) identified and compared alternative designs of marketable water permitting systems. Their work examined the trade-offs among multiple objectives including economic efficiency, equity, ease of implementation and administration, along with environmental concerns. Tisdell and Harrison (1992) proposed a water market procedure using game theory. Their goal was to understand how regulatory agencies could allocate water to promote its equitable distribution. More recently, Winter (1995) provided a review of recent literature addressing the optimal and conjunctive allocation of ground-and surface-water resources.

Conjunctive Use and Irrigation Development

With regard to irrigation water, the implementation of sound conjunctive use projects involves a thorough inventory of soil and water resources and proper zoning of areas suitable for irrigation by surface or groundwater, or where one source can supplement the other. All this requires field surveys and investigations aiming at evaluating hydrometeorological, hydrological and hydrogeological conditions, seepage and soil infiltration rates, crop water requirements and crop patterns, water quality, hydrodynamic parameters and behavior of aquifers, well yields, canal flows and stream discharges, along with the assessment of energy costs to sustain both surface and groundwater development projects. The beneficial effects of conjunctive use in canal commands can be summarized as follows (Karanth, 1987):

- use of groundwater helps cope with peak demands for irrigation and hence reduce size of canals and consequently construction costs;
- supplemental supplies from groundwater bodies ensure proper irrigation scheduling, even if rainfall fails or is delayed;
- groundwater withdrawals lower the water table thus reducing the risk of water-logging, soil salinization and consequent wastage of water for leaching the soils;
- surface and subsurface outflows are minimized, causing reduction in peak runoff;
- when conjunctive use is integrated with artificial recharge the need for lining canals is reduced, as seepage from canals feeds groundwater;
- conjunctive use allows the utilization of saline or brackish ground-or surface-water resources, either by mixing them with freshwater, or by using alternate water resources for irrigation events.

However, there are some constraints that may impair the efficiency of conjunctive use projects, such as:

- increased energy consumption for pumping from wells and for coping with reduction in pump efficiency, due to large fluctuations of water levels;
- administrative difficulties in defining acceptable and equitable groundwater rates, when surface water is available.

RESEARCH THRUST AND DEVELOPMENT

Research needs to be focused more effectively than in the past on planning and management problems of conjunctive use of surface and groundwater. This is the main way to provide planners and decision makers with suitable and well-tested technologies for targeted measures designed to enhance conjunctive use efficiency, while protecting the environment. The lack of research, application of research findings and access to new and advanced technology, is seen as one of the main reasons for the problems plaguing the sector: low efficiency, environmental degradation, high costs and lack of beneficiary responsiveness.

Successful research thrust on sustainable integrated water resource management should include the following actions:

- Data Base Improvement
- Modeling Technology
- Sustainability Criteria
- Spatial Analysis Procedures
- Decision Support Systems

Data Base Improvement

Availability of reliable data on hydro-climatic patterns, water demands, spatial and temporal characteristics of surface and subsurface water bodies is an essential prerequisite for sustainable water resource development.

As long as adequate and reliable data are lacking, planning, design and management of water use projects will remain guesswork, water use irrational and wasteful and development unsustainable. Many projects were conceived and designed on a medium-to long-term basis, on the assumption that future climatic conditions would not change. This might not be so in the years to come, due to global warming and the greenhouse effect.

Therefore, water planners and managers should begin to systematically re-examine engineering design criteria, operating rules, contingency plans and water storage and allocation policies. Demand management and adaptation are essential components for enhancing project flexibility to deal with uncertainties of climate change. In the main, water use planning programs can only be soundly formulated on the basis of adequate data on soil and its productivity, potentially available water resources, water demands, performance of existing water use projects and other related factors.

Modeling Technology

There is no question that modeling technology is already one of the most important and widely used tools for solving long-term environmental problems and will become even more so in the future. Modeling allows to formalize scientific understanding and to integrate the various components and processes involved in the surface/subsurface water system. It also acts as a bridge for knowledge transfer between scientists and policy makers (or the public), as a basis for testing different scenarios and allows to integrate sound planning and management strategies with economic development by means of cost-benefit analyses and finding optimal solutions. Moreover, without modeling technology it would be difficult, if not impossible, to predict the anticipated impacts of any proposed plan and management policy. In the main, the planning, design, management and operation of complex water resource systems depend on modeling technology and its continual development and improvement. Generally, these technologies include simulation and/or optimization models. Since the models embody physical parameters for describing hydrologic systems, bounds on decision and state variables to account for management goals, their reliability depends on the quantity and quality of available data.

Simulation Models

Simulation models are essential for analyzing complex processes of surface and subsurface flow and transport, because they provide a quantitative framework for synthesizing and handling the large set of characteristics that describe the variability of the phenomena, as well as the spatial and temporal trends of hydrologic parameters and stresses, and historical rainfall, flow rate, water level and solute concentration records.

For surface hydrology the most widely used simulation models can be grouped into the following classes (De Vries and Hromadka, 1992), according to the types of hydrologic problems they are intended to deal with:

- single event rainfall-runoff models;
- continuous-stream flow models, accounting in time for precipitation and water movement through the catchment;
- flood hydraulics;
- water quality.

With regard to groundwater hydrology, the two kinds of models most commonly used for solving flow and transport equations are based on finite difference or finite element techniques. The choice between a finite difference and a finite element model depends on the problem to be solved and on user preference. Anderson et al. (1992) provided an exhaustive and still topical literature review on this subject.

Model data requirements vary considerably. In many design studies of surface hydrology, rainfall distribution and volumes are applied to catchments, for which runoff response is characterized by synthetic unit hydrograph and generalized loss-rate functions. For physically based distributed-parameter models, a great deal of data is required for calibration purposes. In groundwater hydrology, knowledge of spatial and temporal characteristics of the aquifers, distribution of piezometric heads, recharge and discharge rates and mass concentrations, along with flow-path information, play a crucial role in testing and validating predictive simulations. The usefulness of forecasting simulations obtained by models is often hampered by the inability to identify the uncertainties and quantify the reliability of model results. Uncertainty in model predictions primarily stems from a number of errors related to the model formulation, such as:

- inadequate concept and description of processes and interactions;
- inadequate description of spatial and temporal variability;
- inadequate description of the state of the system (geometry, initial and boundary conditions, system stresses);

- incorrect parameter identification and improper specification of their error bounds.

As a result, in most cases, the efficiency of a model decreases rather than increases as more and more variables and processes are added. So, one of the greatest challenges for future modeling technology is the development of means by which models can explicitly account for the degree of uncertainty introduced by aggregating the different components.

Optimization Models

One of the major advances in water resource engineering over the last three or four decades, is the development and adoption of optimization techniques for planning, design and management of complex water resource systems. The analysis of these systems may involve thousands of decision variables and constraints. To overcome problems of dimensionality various schemes have been devised, providing decision alternatives which are optimal in some defined sense and which can be used by water managers to assist their decision making.

An extensive literature review of the subject reveals that no general algorithm exists. The choice of method depends on the characteristics of the water system concerned, on the availability of data and on the objectives and constraints specified. Yeh (1985) presented the state of the art and discussed in detail various techniques, mainly for reservoir operations, including linear programming (LP), dynamic programming (DP), non-linear programming (NLP). Combinations of the above methods, along with in-depth analyses, and the merits and limitations of each of them, are also been reported in the literature.

When problems of optimal water resource management include objectives that are difficult to describe due to subjectivity or uncertainty, the principles of fuzzy logic offer a viable approach. The concepts and operational algorithms are available in literature (Zadeh and Kacprzyk, 1992; Parent and Duckstein, 1993; Russel and Campbell, 1996).

More recently neural network approaches (NNAs) have been proposed for solving conjunctive use problems. This technique has been shown to be a powerful tool for both forecasting and optimization purposes, especially when the underlying data-generating processes are uncertain or unknown. Neural network approaches have been used by French et al. (1992) for rainfall forecasting, and by Karunanithi et al (1994) for river flow prediction. Rizzo and Dougherty (1994) used a neural kriging network to characterize aquifer properties, and more recently Raman and Chandramouli (1996) proposed a neural network model for reservoir operating policies.

SUSTAINABILITY CRITERIA

Often conjunctive water use projects involve environmental, economic and social factors that have no numerical scale for assessing their relative importance. Thus scientists and planners have devised various approaches, including weighting, constraints and trade-offs, whereby the above factors may be given assumed values or weights, perhaps based on the findings of opinion polls (Simonovic, 1998). Commonly the issue is comprised under the terms "sustainability" and "sustainable development". These terms are sometimes used interchangeably, but there is a subtle difference between the two. One useful approach is that sustainable development is the process by which we achieve sustainability. According to the Brundtland Commission's Report (WCED, 1987), development is sustainable if it meets the needs of the present without jeopardizing the ability of future generations to meet their own needs.

The concept is inherently holistic: it implies long-term perspective for planning and integrated policies for implementation and improvement. This improvement over time cannot be achieved without rational management of water resource systems: systems able to meet, now and in the future and to the fullest possible extent, society's demands for water and the multiple purposes it serves. The demands will include not only the traditional uses of water flows and storage volumes, that depend on the hydrological pattern of the region, but also the preservation and enhancement of the social, cultural and ecological systems. Considering the above definitions and perspectives, sustainable water resources systems are those designed and managed to fully contribute to the objectives of society, now and in the future, while maintaining their ecological, environmental and hydrological integrity (Loucks and Gladwell, 1999).

This definition completes the one given by the Brundtland Commission, since it links: i) the present needs with society's objectives, ii) the needs of future generations with environmental integrity. Furthermore at river-basin or regional levels, it may not be possible to meet the needs or demands of even the present generation, let alone future ones, if these needs or demands are greater than what the basin or the region can afford. Moreover, degrading the water resource systems will reduce their capacity to meet future needs, whatever those needs will be. So, demand management and degradation prevention play a role of paramount importance in this new concept of sustainable development.

While it may never be possible to determine with certainty whether a water resource development project is sustainable or not, it is possible to devise some procedures that allow to compare performances of alternative options with respect to sustainability. Work conducted on sustainability evaluation has two main emphases (Dorcey, 1992; Duckstein and Parent, 1994, Kroeger and Simonovic, 1997).

The first defines sustainability indicators, as conditions strictly associated with sustainable development, so that their presence is indicative of its existence. The second develops criteria for measuring sustainability such as fairness, reversibility, risk and consensus. Sustainability indicators and/or criteria should be included in the evaluation of water resources engineering projects. The inclusion of such parameters will lead to extending the decision-making process beyond the consideration of purely economic factors alone, toward the development of an evaluation procedure that reflects the new paradigm of sustainability.

The ultimate implementation of such a decision-making framework will require a multi-objective approach to conjunctive use project selection, which must consider the trade-offs between traditional environmental and economic factors and the sustainability parameters. The development of a comprehensive multi-objective decision-making approach that integrates and appropriately considers all these issues, within the water resources project selection process, warrants further research.

Spatial Analysis Procedures

An integrated system approach to developing and testing modeling technology requires different spatial analysis procedures such as: geostatistic methods, Remote Sensing (RS) and Geographic Information Systems (GIS).

Geostatistics, a set of statistical estimation techniques involving quantities which vary over space and time, has found wide application in surface and subsurface hydrology (Matheron, 1971; De Wrachien, 1976; de Marsily, 1986; Kitanidis, 1992). Uncertainty often exists regarding the definition of standards for parameter and input identification for surface and/or groundwater simulation models. This uncertainty often stems from an inadequate understanding of fundamental physical processes, upscaling procedures, methods for integrating and aggregating data in space and time and the spatial and temporal interrelationships of datasets. Geostatistics can help define these standards and provide flexibility for the creation, validation, testing and evaluation of data sets that have distinct temporal and spatial components.

Remote Sensing applications to hydrology are relatively new but are rapidly becoming an important information source for water resources planners and managers (D'Souza and Barret, 1988; Engman, 1992). RS and its continuing advances offer a broad range of techniques for landscape rendering or identifying landscape features, or, in some cases, for actually measuring hydrologic state variables and processes.

Geographic Information Systems are computer systems capable of assembling, storing, manipulating, and displaying geographically referenced information (Dodson, 1992; van Dijk and Bos, 2001). GIS technology appears particularly suitable for dealing with data and modeling issues associated with environments involving multi-scale processes, within a complex and heterogeneous domain. To this end, GIS can help understand spatial data issues and provide a better knowledge of physical phenomena. Most surface and groundwater simulation models are based on finite difference or finite element approximation techniques composed of numerous grid cells. With GIS, distributed model parameters can be regionalized, represented in a digital map and referenced to the modeling grid.

Nowadays, GIS appear to be used mainly as pre-processors for organizing spatially distributed parameters and input data, and as post-processors for displaying and possibly analyzing model results, while GIS-embedded modeling approaches appear to be less common. It is reasonable to hope that with more powerful and affordable computer technology, the integration of GIS with hydrologic models will proceed at a faster pace in the future.

Decision Support Systems

The repeated use of simulation models, linked with spatial analysis procedures, under different hypotheses, whether for system design or operation and management purposes, is generally called Decision Support System (DSS). Decision support systems are interactive computer-based information providers. Like their underlying models and data management components, they do not make decisions.

They merely provide information to those who need it or to potential beneficiaries. DSSs for water resource planning and management provide means of examining the different alternatives involved in the design and management of complex water resource systems. These alternatives regard the extent to which water resource systems can effectively and equitably contribute to the welfare of their users while protecting the environment and increasing the ecosystems' carrying capacity. DSSs can also aid in real-time, adaptive planning and management, where the decisions to be taken, as well as the procedures involved, are continually updated and improved over time.

DSSs are not only analytical tools, but also serve as a means of communication, training, forecasting and experimentation. They can act as links between scientists and decision-makers by providing different scenarios or alternative future environments against which decisions have to be tested. The goal is not to predict the future but rather to learn to live with uncertainty, to factor it into the decision process and to improve the quality of thinking among decision-makers.

In the main Decision Support Systems are suitable for:

- linking simulation and optimization models to determine the values of decision variables or system performance indicators;
- combined use with GIS and other graphic procedures that permit statistical analyses and map displays of spatial data;
- combined use with neural networks able to learn to reproduce results of complex physical and chemical processes and hence provide "black boxes" for the above processes.

FRAMEWORKS FOR IMPLEMENTATION OF CONJUNCTIVE USE R & D

Institutional strengthening and proper financial assessment are essential tools for the rational planning, design and management of water resource systems. On the one hand, without a sound institutional framework at the watershed or regional level, it will not be possible to promote and ensure efficient integrated water resource management. On the other, economic constraints are equally important. The costs of system improvements are normally tremendous and governments, in this era of transition from state to market, will no longer be in a position to finance water planning, design and management activities. This new philosophy is based on the "user pays" principle.

So, research thrust and sustainable development, as defined earlier, should meet two basic requirements, namely institutional strengthening and economic viability. To this end, appropriate institutional and financial analysis in all phases of water planning projects, is a prerequisite for the sustainable and conjunctive use of surface and groundwater.

Institutional Issues

In many countries the close of the twentieth century was marked by the change in the global economy from state ownership to a market oriented economy. The immediate outcome has been the move from subsidized towards privatized projects in the water sector.

On the one hand, privatization implies reduced government financing in water projects, on the other the move towards a market economy calls for improvements in both services and goods supplied by the projects. Added to this is the increasing public awareness of environmental issues. As a result water resources development in the twenty-first century will be obliged to:

- rely to a lesser extent on government funding;
- provide a higher level of service;
- abate soil, air and water pollution.

In this context, operation, management and maintenance of water resource systems should also be improved. Governments will cease subsidizing water resources use and planning with the result that, the main beneficiaries, i.e. the users, will ultimately take over the government role of providing basic services. As individuals they can do little, but as water users' groups and associations, they will probably be able to do a better job. than bureaucratic frameworks.

Obviously, the transition from state owned enterprises to a market oriented economy will be a lengthy and slow process in which each party will try its best to defend its ideas and interests. Moreover, the laws and rules regulating different aspects of water use and related subjects are outdated and often obsolete.

If legislation and regulations are not sufficiently dynamic to accommodate the rapid changes taking place in society, they become obsolete and are no longer able to reflect the changing socio-economic conditions of groups and individuals. Consequently, one of the most important issues to be addressed is the conceptualization of the form of institutional and financial interrelationships in water resources management so as to clarify stakeholders' responsibilities and avoid conflicts. In the strive to face these challenges, research and expertise should focus on:

- the role of governmental institutions and the private sector in maintaining effectiveness of water use and planning projects;
- the processes by which the level of services should be redefined and specified;
- how the costs of redevelopment and of future operation and maintenance, should be determined, recovered or shared to assure agreed service levels.

Financial Issues

Operation of water resource systems requires an agency capable of managing surface and groundwater flows and levels. If the system is to be financially sustainable the costs of developing and managing water resources will need to be recovered from the beneficiaries or from the community.

To link all the financial aspects involved in water resource systems, a conceptual framework is needed, based on the "user pays" principle (van Hofwegen, 1997). This entails identifying beneficiaries or clients for the services provided. The clients need to be clearly identified so as to establish with whom to enter into a service agreement, who to charge and where to send the bill.

Levels of service are based on operational standards and quality criteria.

Operational standards regulate system management and serve two purposes:

- to establish criteria for operational performance measurement;
- to establish criteria for service delivery.

The quality of service provision can be defined as a combination of parameters relating to the adequacy, flexibility, convenience, cost and security of the services provided. To assure service delivery requires the design, construction, operation, maintenance and eventual replacement of a hydraulic infrastructure. The costs of managing this infrastructure can be broken down as follows:

- operation costs, i. e. the human resources required for operating the system as a whole;
- maintenance costs, determined by the number of control units and their individual costs;
- depreciation costs, which reflect the age, wear and tear of the infrastructure.

Improving service levels will automatically entail increase in management ability or upgrading of the infrastructure or both. So an acceptable trade-off should be found between investments and management ability to achieve a certain service level, assuming service level to be one of the management goals.

Clearly defined service level agreements and effective accountability systems are the basic requirements for a reliable and effective service. In this way a direct relationship is established between demand, level and cost of service on the one hand, and payment for services on the other. To minimize cost variations, hence the service charges, proper planning of service cost is essential. In this context, a reliable asset management plan is an indispensable tool.

CHALLENGES FOR THE FUTURE

At the beginning of the twenty-first century, conjunctive use of surface and groundwater is coming under pressure on a number of fronts. The expected demand for water exceeds available resources, plans fall short of targets, population is increasing, though growth rates are slowing down, and economic crises coupled with environmental concerns further complicate and exacerbate efforts under way to tackle these problems. The scientific and professional communities recognize the causes and effects of the problem. These, in turn, create a number of challenges for the future, which can be summarized as follows (IAHR, 1999):

- Rainfall/Runoff processes and modeling;

- Groundwater management, monitoring and remediation;
- Unsaturated zone: exchange with surface water and coupled flow;
- Conflict resolution in water management;
- Water resource management under climate change.

Rainfall/Runoff Processes and Modeling

The quality and performance of rainfall/runoff models have improved but there is still room for further improvement. Future research directions will include adaptation of models for use in the domain of hydroinformatics, refinement of flow and quality modeling, examination of management options (storage and treatment facilities), interactive, dynamic control required in real-time simulation studies.

Groundwater Management, Monitoring and Remediation

Today problems in groundwater management are concerned with overexploitation, water table lowering, water deficit and pollution. Improved land management, to increase groundwater recharge by reducing evaporation and, where appropriate, runoff, also warrants investigation.

Further problems are created by land subsidence due to over exploitation. Any groundwater management activity has to be based on an adequate and thorough field investigation, calling for improved methods in this area.

Accurate monitoring of the groundwater flow and quality, including estimate of techniques discharges and storage is also important. Restoring the quality of polluted groundwater entails not only the elimination of pollution sources but also the remediation of contaminated groundwater in both the saturated and unsaturated zones.

The mechanisms of multiphase flow, mixing and mass transfer in porous media are not sufficiently understood and the efficiency of different clean-up techniques has not been fully assessed.

Central to this understanding are tremendous uncertainties that plague groundwater pollution and remediation problems.

Dynamic and stochastic simulation models of water flow and solute transport in saturated and unsaturated zones, combined with carefully selected field experiments, are crucial tools for proper assessment and management of groundwater contamination, and need to be further improved.

Unsaturated Zone: Exchange with Surface Water and Coupled Flow

The zone of groundwater level fluctuations is characterized by frequent changes between saturated and unsaturated conditions. To describe and control the processes occurring within this zone, a major effort has been launched during the last two or three decades by scientists of different disciplines. Unfortunately, segmented disciplinary research has hindered up to now, an exhaustive experimental and theoretical understanding of the unsaturated zone.

So, challenges for both basic and applied research remain, as does the urgency for progress inasmuch as the soil and groundwater resources are increasingly vulnerable to long-term pollution. Knowledge of the interactions, feedbacks, biochemical and hydrochemical processes is far from sufficient in many key areas.

To this end, it is essential to couple chemical and biochemical phenomena with flow and transport processes. The kinetics of the reactions must be better controlled by the transport mechanism that brings the reactants into contact with one another. It is therefore essential to understand the exact coupling of flow and transport and the bio-geochemical mechanisms. Another theme strictly related to behavior in the unsaturated zone concerns flow and transport phenomena in the soil-plant-atmosphere system.

A new global approach is needed in order to achieve a better understanding of the behavior and evolution of the unsaturated zone and to develop suitable strategies for its optimal conservation, use and management.

Conflict resolution in Water Management

Water, an increasingly scarce resource, is already a source of conflict and likely to become even more so. In this context, scientists are faced with the challenge of developing suitable methodologies for emergency management and conflict resolution strategies. In this regard, simulation models linked with programming techniques for optimal management are crucial tools for proper planning and design of efficient and equitable integrated surface and groundwater use projects.

Technology and expertise transfer from the scientific community to practitioners should be encouraged and enhanced Decision makers also need to be able to recognize the potential of problem-solving measures, including emergency management and conflicts among different water uses and users.

To this end, procedures for supporting negotiations for water allocation should be developed and tested for international, national or watershed applications. A methodological framework for comprehensive environmental risk and impact assessment also needs to be developed to assess environmental vulnerability and resilience.

Water Resource Management under Climate Change

The ongoing debate about climate change has raised questions about possible effects of such a change on the hydrologic environment.

According to the Intergovernmental Panel of Climate Change (IPCC, 1996), the planet warming will affect precipitation patterns, evapotranspiration rates, the timing and magnitude of runoff, the frequency and intensity of storms and, therefore, both the quality and quantity of surface and groundwater resources.

In addition temperature and precipitation changes will affect the water demands for agricultural, municipal and industrial purposes. Although climate change is expected to have significant impacts on water resource systems, the degree of uncertainty remains significant, and the adequacy of existing water planning principles and evaluation criteria in the light of these potential changes is unclear.

So, research needs to be encouraged to improve the scientific knowledge about the relationships between climate patterns, land use and the hydrologic systems for different climatic settings. This, in turn, will help water planners and managers to reassess operational principles currently being applied.

CONCLUDING REMARKS

The world's fresh water resources are unevenly distributed in time and space. Until recently water resource management focused mainly on reallocating water when and where it was required, a supply-side or fragmented approach.

Nowadays, it is apparent that water availability is dwindling due to both population growth and increased per capita water use causing often irrepairable damage to the environment.

To face this challenge a new holistic, systemic approach, relying on conjunctive use of surface and ground water resources is needed to overcome the current fragmented management of water. This implies long-term planning and management strategies with respect to both water quantity and quality.

To ensure sustainability, water resource systems need to be planned, designed and managed is such a way as to fully meet the social objectives of both present and future generations, while maintaining their ecological, environmental, and hydrological integrity. This imposes constraints on every stage of development from project planning to final operation and management.

This new holistic approach entails system analysis and modeling, which requires the identification, analysis and evaluation of the interactions between all the components of water resource systems over space and time.

These components should be integrated into a network or system involving relationships between humans and their institutions, nature and technology. For such systems to be sustainable, they must interact smoothly with other social subsystems and adapt to changes and uncertainties in supply and demands.

Multiple alternatives should be defined and evaluated with respect to overall system performance objectives. Managers and decision-makers have to consider a large number of often conflicting demands on the available water, and develop and operate water resource systems under numerous social, economic and legal, as well as physical constraints.

The various interests at stake call for a decision process involving multiple objectives, multiple users and multiple constituencies and stakeholders. In this context, the traditional approach whereby water resources development and management was a "government's business" needs to be replaced by a participatory approach involving both governments and stakeholders at all levels.

Experience shows that stakeholder participation must be genuine and not symbolic, and that user associations must have a decisive role in the decision process as to what is done, how it is done and who pays for it. Experience also shows that partnerships between governments and stakeholders can be effective with governments playing a vital role in creating an enabling environment and in providing technical support and research thrust.

Economic constraints are equally important in water resource development. The cost of water system operation and improvement is normally tremendous, and governments, in this era of transition towards a market-oriented economy, will be unable to continue financing activities as they used to.

The new philosophy is based on the principle that the services must be paid for by those who benefit from them. Sustainable development, as previously defined, requires a sound financial management framework in which the revenues from service provision cover the costs.

In the strive for sustainable development in water resource use and management, the effectiveness of any instrument devised to realize that goal, depends ultimately on two factors: the approach adopted and the quality and number of the individuals responsible for pursuing that approach.

On the one hand, only the holistic and systemic approach outlined herein seems capable of providing a deeper understanding of the behavior and evolution of surface and groundwater systems. This approach will also design strategies for maximizing the efficient use and management of available water resources, while preserving or enhancing the buffering capacity of the environment against unexpected stress or negative long-term trends. On the other hand, engineers, economists, ecologists, planners, along with stakeholders, user associations and all sectors of society must be involved in the decision making process. Failure to do so will increase the risk of opposition to projects even once they have been designed or constructed. Successful collaboration with an informed and attentive public can lead to more socially compatible uses of water resources and to more creative, appropriate, and hence sustainable uses of technology for addressing community or regional water resource problems or needs.

A GLOBAL PERSPECTIVE ON WATER SCARCITY AND POVERTY

The Picture in Brief

- Irrigation has played a major role historically in poverty alleviation by providing food security, protection against famine, and expanded opportunities for employment both on and off the farm. Development of irrigated agriculture has been a major engine for economic growth and poverty reduction.
- One of the most recent examples of success in poverty alleviation has been in eastern India and Bangladesh. Here the spread of low-cost tube-well irrigation has stimulated growth in agricultural production and contributed to the reduction of the number of people below the poverty line.
- The growing scarcity and competition for water, however, stands as a major threat to future advances in poverty alleviation. Food production is likely to be adversely affected particularly in the semiarid regions, which include two of Asia's major breadbaskets, the Punjab and the North China Plain. An increasing number of the rural poor are coming to see entitlement and access

to water for food production and for domestic purposes as a more critical problem than access to primary health care and education.

- o The development of tube-well irrigation has contributed significantly to the increase in food production and reduction in poverty. However, in many of the more arid regions the overexploitation of groundwater poses a major threat to environment, health, and food security-a threat to the welfare of the poor far more serious than that posed by the widely criticized construction of large dams.
- o Poverty persists in many of the rain-fed and upland areas, the so-called marginal areas most of which can be described as water scarce. Much of sub-Saharan Africa, with its lack of developed water resources, fits this category. Despite the apparently wide range of potentially useful technologies, efforts to increase productivity and alleviate poverty have thus far met with limited success.

Some Challenges

- o In an environment of growing scarcity and competition for water, a comprehensive strategy is needed to improve the productivity of water in both irrigated and rain-fed agriculture, and to ensure access to water by poor men and women.
- o As water is withdrawn from agriculture, in the management of irrigation systems, more attention must be given to water needs for multiple uses not only for agriculture, but for other domestic uses, and for environmental needs.
- o Policies and institutions must be developed and cost-effective management practices adopted to halt the environmental degradation caused by overexploitation of groundwater resources.
- o A basin approach to management and allocation of groundwater and surface water will require new

institutional arrangements to assure both efficiency and equity in the use of basin water resources and to protect the interests of the poor.

- Special attention must be given to implementing policies and developing technologies suitable for adoption by resource-poor farmers in water-scarce or marginal upland and rain-fed areas, particularly those in sub-Saharan Africa, as well as those in well-watered areas such as eastern India.

The Issue

Poverty eradication through sustainable development can be regarded today as perhaps the central goal of the Consultative Group on International Agricultural Research (CGIAR), of most agricultural research and development institutions in the developing world, and of the national governments that support their research. Irrigation has played a central role in poverty reduction in the past. But the growing scarcity and competition for water and the overexploitation of groundwater resources are putting the poor in irrigated areas at great risk. In addressing the poverty problem, we must consider the impact of reduced water availability for irrigation not only on crop production, but also on the wide range of other uses that are a part of the livelihood of rural agricultural communities. Meanwhile, poverty persists in many of the rain-fed and upland areas, the so-called marginal or traditionally water-scarce areas. In these areas, the inability to effectively mobilize water resources has prevented farmers from using modern yield-increasing inputs and raising incomes.

There are two regions of the world that stand out in terms of the scope and magnitude of rural poverty-South Asia and sub-Saharan Africa. They could not be more contrasting in terms of water resources and irrigation development and hence the challenge posed to International Water Management Institute (IWMI) scientists and others for poverty alleviation. In South Asia, close to 50 percent of the cereal grain area is irrigated. Two-thirds of cereal grain production and most of the marketed surplus comes from the irrigated areas. On the contrary, in sub-Saharan Africa, the contribution of irrigation to cereal grain production is about 5 percent.

In this paper, we discuss the implications for poverty alleviation of growing water scarcity with particular reference to South Asia and sub-Saharan Africa. First, we briefly summarize the impact of irrigation development on poverty alleviation in South Asia in the recent past. Emphasis is placed on the role that advances in irrigation technologies and private development of groundwater resources have played in providing the poor access to water.

Then, in an environment of growing water scarcity, we examine the challenges that lie ahead for the development of water resources leading to sustained poverty alleviation. Between the 1960s and the 1990s, real food grain prices fell by approximately 50 percent. This decline was principally but not entirely due to the impact of the so-called green revolution in the developing countries.

The subsidization of food grain production by the developed economies also contributed to the decline. Determining the precise share of the gains in cereal grain production attributable to new seeds, fertilizer, and irrigation is an almost impossible task. However, there is little doubt that without the advances in irrigation technologies and extraordinary investments in irrigation expansion by both the public and private sectors, the impact of the green revolution would have been greatly reduced.

The benefits of lower food grain prices to the poor are obvious. Sixty percent of the money spent on food by the people below the poverty line in Asia is apportioned for cereals (which provide as much as 70 percent of their total nutrients). A second major impact of irrigation is in the employment generated both on and off the farm, providing entitlement or purchasing power for the poor. For landless laborers, increased cropping intensity has the greatest impact on employment. Irrigation means more work in more days of the year. The employment impact is felt not only in irrigated areas but also in rain-fed areas. Sometimes, landless workers in rain-fed villages migrate long distances to take advantage of employment opportunities in the irrigated areas. There is also a multiplier effect, with higher incomes in agriculture creating employment opportunities off the farm.

Irrigation and Poverty Alleviation-Past Achievements

Between the mid-1970s and 1990, the number of people below the poverty line in India fell from over 50 percent to approximately 35 percent (Datt 1998) while in Bangladesh it fell significantly (Palmer-Jones 1992). However, the absolute number of people below the poverty line increased slightly. An important contributor to poverty alleviation was the growth in public-sector-funded canal irrigation and in largely private-sector-funded tube-well irrigation (Shah 1993). Technological advances in storage dam construction and in tube wells and lift pumps enhanced this expansion and allowed a larger portion of the crops to be grown during the drier periods of the year thus avoiding crop losses due to flooding and typhoons. In addition, the dry season environment of high solar energy and low pest infestation gives the highest crop yield response to fertilizer application.

Tube-well technologies in particular were instrumental in poverty reduction in eastern India and Bangladesh, where agricultural development was long seen as an intractable problem. The initial emphasis in the 1960s and1970s was on publicly financed deep tube wells with command areas of around 20 hectares. These systems performed poorly and management committees tended to be dominated by the richer and influential farmers (Boyce 1987). The gradual expansion of private tube wells, often within the deep tube-well command areas, also favored the elite who had access to the required capital (Shah 2000). However, three important developments have allowed small farmers and poorer farm households to gain access to irrigation water:

(i) the development of water markets,.

(ii) the introduction of cheaper and smaller Chinese engines, and

(iii) the widespread adoption of treadle pumps (Palmer-Jones 1992, Shah et al. 2000 and Shah 2000).

For those that purchase water, there is strong pressure to economize on groundwater use, whereas tube-well owners feel no such pressure. The manually operated treadle pump, which requires an investment of less than Rs700 (US$18), is particularly well suited to farmers with less than an acre of land. To facilitate the spread of these technologies among the small farmers, credit schemes were developed by the Grameen Bank and other NGOs with particular emphasis being given to the role of women in obtaining loans and managing the pumps (van Koppen and Mahmud 1996).

While eastern India and Bangladesh have achieved a great deal of success in developing water resources to help attack poverty, many problems still remain. Particularly in India, major public policy initiatives have actually impeded groundwater development rather than expediting it (Shah 2000). Pump subsidy schemes need to be reformed. Efforts are also needed to promote a wider variety of diesel and manual pumps, to improve the fuel efficiency of pumps,and to experiment with innovative measures for fee collection for use of electricity. The potential for further development of groundwater in eastern India is great. Less than a quarter of the groundwater resource is in use at present.

IMPACT OF WATER SCARCITY ON THE POOR

As we approach the next century, it is widely recognized that many countries (in contrast to eastern India and Bangladesh) are entering an era of severe water shortage. IWMI has undertaken a long-term program to improve the conceptual and empirical basis for the analysis of water in major river basins of the world (Seckler, Molden, and Barker 1998). The initial findings of this study project that in the first quarter of the next century 2.7 billion people or a third of the world's population will experience severe water scarcity. The bulk of this population will reside in the semiarid regions of Asia and in sub-Saharan Africa. Due to overexploitation of groundwater, food production will be adversely affected in the semiarid regions, which include two of Asia's major breadbaskets-the Punjab and the North China Plain. In northwestern India, the rise of tube-well irrigation crowded out manual water-lifting devices used by smallholders during the 1960s. In western and peninsular India, besides competitive lowering of groundwater tables, excessive pump irrigation has also resulted in fluoride contamination of groundwater, which has been the mainstay of the poor for their domestic water needs.

What are the implications of these findings for the poor? Water is both a commodity and a natural resource and a perceived human entitlement. When Nobel Laureate, Amartya Sen (1981) wrote about poverty and famines in Bengal, he spoke of "entitlements" in terms of purchasing power for food. The primary people affected by the famines were the landless rural poor. But in today's environment of growing water scarcity the problem is more pervasive. An increasing number of the poor-rural and urban consumers, rural producers, and rural laborers-are coming to view access or entitlement to water as a more critical problem than access to food, primary health care, and education.

The typical urban household uses water for drinking and sanitation. But rural areas use water for a wide range of purposes. Even in irrigated areas water is used not only for the main field crops but also for domestic use, home gardens, trees and other permanent vegetation, and livestock (Bakker et al. 1999). Other productive uses include fishing, harvesting of aquatic plants and animals, and a variety of other enterprises such as brick making. In addition, irrigation systems can have a positive or negative effect on the environment. Thus, the withdrawal of water affects the rural household, rural economy, and environment in a number of ways. Water scarcity is exemplified by situations such as: the need to carry heavy pots of water several kilometers every day to meet household needs; the destitution of farmers who lose their lands or of the landless who lose their jobs because of lack of irrigation water; the loss of wetlands or estuaries because of upstream water depletion; and increasing health problems due to water pollution and a rise in incidence of water-borne diseases.

Experts in the field agree that the quantity of water is even more important than the quality in terms of its impact on human health. However, water scarcity leads to declining water quality and pollution, which has an especially adverse impact on the poor. Many (perhaps most) of the poorest people in developing countries are forced to drink water that is unfit for human consumption.They suffer from a range of skin-related and other health problems.

As water is withdrawn from agriculture, in the management of irrigation systems, more attention must be given to water needs for domestic and health purposes, and to other consequences such as the impact on the environment. Unfortunately, in the case of water for agriculture, allocation cannot be accomplished solely through pricing mechanisms. Along with farmers other stakeholders, including the poor, should have a voice in how limited water supplies are to be allocated.

EMERGING GROUNDWATER PROBLEM

As noted above, the development and expansion of tube-well irrigation contributed significantly to the increase in food production and reduction in poverty. However, in arid and semiarid regions, the point has now been reached where the overexploitation of groundwater poses a major threat to environment, health, and food security-a threat to the welfare of the poor far more serious than that posed by the widely criticized construction of large dams.

The regions of China, India, the Middle East, and north Africa, projected in an IWMI study to face severe water shortage, are the areas where overexploitation of groundwater has been most pervasive. These areas have had a "free ride" over the past 2-3 decades. Tube-well yields attributable to irrigation have typically been more than double the yields from canals and tanks. But the penalty for mismanagement of this common-pool resourcc is now coming due.

The groundwater problem has two contradictory aspects. First, there is the rapid drawdown of freshwater aquifers mainly due to the worldwide explosion in the use of wells and pumps for irrigation and for domestic and industrial water supplies. Second, there is the opposite problem of rising water tables of saline and sodic water. Added to these two problems is the pollution of aquifers by toxic elements. Here again the poor often pay the price not only through loss of crop production or cropland but also through a shortage of water for the range of uses noted above and through increased health problems. The recent discovery of extensive arsenic poisoning of the aquifers in Bangladesh and fluoride contamination of well water in western India provides a prime example of the adverse effects of groundwater mismanagement in some of the poorest countries of the world. While the problems of groundwater are clear, the solutions are not. The pricing and regulation of pumping of this common-pool resource are not feasible. Even if technical and management capacities are available, which country would be willing to pay the enormous cost of this policy in terms of reduced food production and domestic and industrial water supplies? The best way to recharge aquifers needs more careful analysis. Much of the seepage and surface runoff from canals recharges groundwater aquifers.

Seen in this light, Chambers (1988) suggests that a major and perhaps the main beneficial effect of new canal irrigation is to distribute water through the command allowing it to seep and so provide water for lift irrigation. Dhawan (1993) points out that according to his calculations half of the crop output originating from tube-well irrigated lands in the Punjab is from groundwater that is of canal origin. In short, much of South Asian irrigation must be analyzed and should be managed from a basin perspective taking into account the complementary relationship between canal water and groundwater. While there are technical solutions for reducing salinity, the financial costs of various options and the appropriate procedures for implementing management strategies at farm and system levels are not well understood. In short, given the truly alarming threat of fresh groundwater depletion in the world, it is astonishing how little attention, whether in research or action, is given to this problem.

Marginal Areas

Poverty persists in the irrigated areas but especially in many of the rain-fed and upland or the so-called marginal areas. In fact, most of the areas of persistent poverty are areas that can be described as water-scarce. Much of sub-Saharan Africa, with its lack of developed water resources, fits into this category.

There is a striking dissimilarity between the physical and socioeconomic environments in the irrigated lowlands of Asia and the rain-fed, water-scarce areas in sub-Saharan Africa and Asia. Strategies for water management that have been successful in reducing poverty in the former environment have proven to be largely inappropriate for the latter. There appears to be relatively few lessons from the South Asian experience with irrigation and poverty reduction that can be readily transferred to water-scarce rain-fed areas in sub-Saharan Africa and elsewhere. In fact, little progress has been made in raising agricultural productivity even in the water-scarce rain-fed areas of Asia. In these less favorable environments, the road to increased productivity may lie not only with the introduction of new technologies but also with the exploitation of traditional technologies (Agarwal and Narain 1997).

To date, efforts to mobilize water resources following conventional canal irrigation system designs have been largely unsuccessful. Carruthers and Clark (1981) cautioned that without careful pre-appraisal "the farmer's promised banquet may turn out to be simply an engineering picnic." And indeed many of the early irrigation projects in Africa all too frequently turned out to be "engineering picnics." Jones (1995) reports that 18 World Bank irrigation projects in sub-Saharan Africa cost US$31,238 per hectare, while 56 South Asian projects cost $1,746 per hectare. This explains why such a large portion of African projects received an "unsatisfactory" rating. Furthermore, many of these systems were in locations where either the physical or the socioeconomic environment, or both were ill-suited to irrigated agriculture. Many scientists strongly believe that the rain-fed areas and natural wetlands offer the greatest potential for production gains in the immediate future. Scientists have shown that a number of water harvesting and supplemental irrigation technologies hold greater promise for increasing crop yields. But their adoption by farmers has been extremely limited, as the risk and costs seem to have outweighed the benefits (Oweis, Hachum, and Kijne 1999). In sub-Saharan Africa, part of the problem may lie in the much higher cost.

For example, a major difference between South Asia and sub-Saharan Africa is the cost of pumps and irrigation equipment. Because of the rapid and large-scale groundwater development, South Asia developed its own highly competitive pump industry that delivers low-cost pumps to small farmers for which, in Africa, farmers have to pay much more. This may also be an important reason why even smallholder irrigation development in sub-Saharan Africa region has occurred only with massive government and donor support.

Developing a more intensive agricultural system will prove more costly in sub-Saharan Africa than in Asia for three reasons. First, as noted above, the equipment will be more expensive. Second, and related to the first, as IWMI's research in Burkina Faso and Niger emphasizes, there has been little time for the development of institutions such as credit, marketing, land and water rights that are essential to the sustainability of intensive agricultural systems (Abernethy et al. 2000). One cannot simply rely on the market to bring about the institutional changes needed. A strong role must be played by public policy and the state in facilitating the institutional changes and adoption of technologies that will increase agricultural productivity (Lele and Stone 1989; Binswanger and Pingali 1988). Finally, the ecological risks associated with intensification, particularly pests and diseases, might be more serious than those in Asia because intensive cropping has had no time to co-evolve with the natural biotic environment.

CHALLENGE FOR POVERTY ALLEVIATION

With the onset of the population explosion following World War II, poverty in the developing world has been associated with hunger and food insecurity. The response has been to produce more food by increasing crop yields and expanding irrigation. Despite the advances in global food security, it has become increasingly clear that productivity alone could not win the war against hunger. Poverty alleviation has become associated less with food production per se and more with livelihood (Chambers 1988), with employment, and with Amartya Sen's entitlements to food (Sen 1981). Identifying a poverty line based on daily per capita income and measuring the reduction in the number of people below the poverty line have constituted the popular "yardstick" by which progress in poverty alleviation is measured. The growing scarcity of water relative to food in many parts of the world suggests that this may no longer be an appropriate yardstick. Access or entitlement to water is not simply a matter of having more money or purchasing power. Water security is an increasingly important element of any poverty eradication program.

Much is known already about policy and program design for food security and even about policy frameworks required for environmental protection.

By contrast, little is known about the appropriate mixes of policies, institutions, and technologies that could help achieve water security for both men and women in water-stressed environments (Webb and Iskandarani 1998). Furthermore, because of multiple use of water, there are a large number of stakeholders among government agencies and the private sector who have a keen interest in water allocations. A basin approach to water management and allocation will be required to utilize water more efficiently (Seckler 1999) and to take into account off-site effects of more narrowly focused projects such as reforestation, which can have a negative impact on marginal upland farmers (Finlayson 1998; Starkloff 1998). As water becomes scarcer, the conflict over water allocations, rights and entitlements at farm, system, and basin levels is bound to increase. New water basin institutional arrangements will be required to assure both efficiency and equity in the use of basin water resources and to protect the interests of the poor.

Irrigation consumes about 70 percent of the world's available water. Subsidies for infrastructure have been reduced, the most accessible and cheapest water resources have been developed, and in an increasing number of basins all of the water has been committed. Demand for water for higher-valued uses-domestic, industry, and hydropower-is rising. The agricultural sector must produce more food with less water. Achieving this goal is thought by many to lie largely in improving the efficiency of canal irrigation systems. As suggested above, however, we believe that there must be more attention through research and action programs on the management of groundwater and surface water in a basin context in irrigated areas and on assessing the potential of alternative low-cost micro irrigation technologies in raising crop production in water-scarce rain-fed areas. There is an urgent need for new strategies to improve the productivity of water in both irrigated and rain-fed agriculture, and ensure access to water and technologies by the poor. As new policies, institutions, and technologies are devised and tested to achieve this goal, a major task of IWMI will be to design strategies to enhance water security for poor men and women, a critical step in poverty eradication.

FOCUS ON GROUNDWATER RECHARGE

It is impossible to overstate the cruciality of need of water for man's civilised living, indeed his very existence. It is the most precious gift of nature and fortunately to LIS living in India's geographical boundary very plentiful. A combination of 6000 kin. long shoreline, 2500 km long unbroken chain of high Mountains in the north and north-east, a very long and hot summer creates a giant precipitation-making machinery resulting in India's average annual precipitation of 543 min, which is greater than global average of 314 mm.

This Sustains an annually replenishable utilisable surface water resources of 690 km3 and ground water resources of 420 km^3, In totality these resources can meet India's freshwater need, 1050 km^3, estimated up to 2025 AD. So there is no cause for immediate worry for a quarter of a century, superficially. But there is cause for worry, firstly due to space-time variation in rainfall and consequent runoff from west to far east (almost nil rainfall in Thar desert to 11 in Meghalaya) and it's activity confined to only four (June to September) watery months. Then, there are acute variations year after year, giving rise to drought and famine Situations as also flood revages.

As a corrective measures a lot has been done in terms of creating storages (to obviate time variation) and canal construction (to obivate space variation) in preceding half a century since Independence. The interaction of surface flow and groundwater flow has prompted the planners to take major initiative to plan on the basis of conjunctive use, i.e. surface and groundwater taken together, and not to treat them separately. This is possible only with improved recharge methods and pumping.

Conjunctive use of Surface and Ground Waters

Bulk of surface and groundwater flows, about 75%, are used in irrigation, but still greater part of country depends on rain-fed agriculture, Failure of monsoon exposes the agriculture to drought and famine and other damaging effects, while above-average monsoon results in flooding, inundation and resulting devastation. Conjunctive use combines the advantage of groundwater storage and surface water systems and leads to efficient water management.

National Water Policy 1987 document directs that both surface and groundwater should be viewed as a integrated resource and should be developed conjuctively and in a coordinated manner. In times and places of higher rainfall or high availability of water on surface its transfer to groundwater can be induced by artificial recharge methods while during low rainfall period groundwater can augment surface supplies. Conjunctive use permits better flexibility in cropping pattern, helps in blending of brackish groundwater with fresh water in and areas.

GROUNDWATER POSSIBILITIES IN INDIA

There are three type of formations in India that bear and yield groundwater. A large portion, about 36% area, consists of consolidated hard rock formation with little pore space but having joints, fractures and shear planes which can yield water by dug well or bore well upto as little as I lit. sec. In semi-consolidated formation, about 1.5% area that includes Rajasthan, there is good porosity and tubewells and deep dug wells yield about 40 lit./sec. At village Borunda about 105 m from Jodhpur dugwell encounters water at 50 in depth arid yields nearly 500 lit/sec without showing any drawdown. In vast areas covering unconsolidated formations in Indus, Ganga, Brahmputra, and coastal basins vast potential exists for dug, bore and tubewells to be added to vast number that already exists due to good porosity arid availability of water to recharge arid replenish the pumped or drawn out water. However. the over exploitation of groundwater should be avoided and artificial recharge should be encouraged there.

ARTIFICIAL RECHARGE OF GROUNDWATER

Recharge of groundwater (GW) is a component process of nature's hydrologic cycle in which precipitation incident on collected on ground surface percolated through it to augment the groundwater storage. When this moment from Surface to underground formation needs to be augmented beyond the natural process it is termed 'artificial recharge'. The latter are designedtoserve one or more of the following purposes:

I. To maintain or augment natural GW as an economic resource.

II. To combat adverse conditions Such as progressive lowering of groundwater levels, unfavourable salt balance and (coastal) saline water intrusion.

III. To store underground the surface waters imported from other basins.

IV. To provide treatment and storage for reclaimed wastewater for possible reuse.

Besides the obvious advantages of purposes listed in 1, 11 and III above, the last one has a particular relevance in Western Europe and USA.

A variety of methods have been developed to artifically recharge the groundwater choice of which is governed by local topographic, geologic and soil conditions. the quantity of water to be recharged, and ultimate water used intended. They are primarily based on water spreading, and classified as basin, stream channel, ditch and furrow, flooding and irrigation methods, apart from use of recharge wells. Hydraulics of a recharge well is alike that Of pumping well with direction of flow reversed, superficially speaking.

However, recharge rates seldom equal pumping rates for other conditions being same, due to air and fine material carried by recharging water often partially clogging the pore space and reducing acquifier permeability. However, recharge wells are convenient for disposal of septic tank effluent, excess irrigation water. The recharge wells have been shown to be effective in combating GW salinity conditions, particularly in coastal regions.

RESEARCH FINDINGS AND LATEST DEVELOPMENTS

Research has continued on spreading methods, particularly basin and stream channel methods, to increase efficiency of (artificial) recharge. Continued recharge shows gradual decrease of recharge rates with time which is attributed to microbial growths clogging the soil pores. This has been confirmed by experimenting on sterile soils which show no decrease with time. This growth can be checked by addition of organic matter or chemical soil conditioners, or alternate wetting and drying of soil. Drying is believed to kill microbial growth and opens soil pores. Effect of other factors like sunlight, temperature and algal growths in water is under investigation.

Nuclear well logging techniques can be used to identify porosity of underground stratums. Methods using natural and artificial radioisotope tracers have been reported for pre-storage investigations to delineate the storage areas for purposes of artificial recharge. Nuclear techniques can be used to find if a GW body is getting replenished, if so the rate of recharge.

Bibliography

Bailey, Conner and Charles E. Faupel: *Environmentalism and Civil Rights in Sumter County*, Colorado, Westview Press, 1992.

Bailey, Ronald: *Liberation Biology: The Scientific and Moral Case for the Biotech Revolution of Water*, Amherst, Prometheus Books, 2005.

Barnum, Susan R.: *Biotechnology: An Introduction to Wastewater*, Belmont, Thomson/Brooks/Cole, 2005.

Campbell, R. R., and Wade, J. L.: *Society and Environment: Treatment of Water Technology*, Boston, Allyn and Bacon, 1972.

Chrispeels, Maarten : *Plants, Genes and Water*, Sudbury MA, Jones and Barlett Publishers, 2003.

Dodds, John H.: *Plant Genetic Engineering and Wasteater*, New York, Cambridge University Press, 1985.

Gelobter, Michael: *The Distribution of Air Pollution by Income and Race*, Berkeley, University of California, 1989.

Kruvant, W.J.: *People, Energy, and Water Pollution*, Cambridge, Ballinger Publishing, 1975.

Retzer, W.J.: *Biotechnology Workbook and Water Treatment*, Englewood Cliffs, Prentice Hall, 1991.

Vallette, J.: *The International Trade in Wastes: A Greenpeace Inventory*, Washington, Greenpeace, 1989.

Walden, Richard: *Genetic Transformation in Plants and Wastewater*, England, Open University Press, 1988.

Zupan, J.M.: *The Distribution of Air Quality in the New York Region*, Baltimore, Johns Hopkins University Press, 1973.

Index